I0516068

tredition

tredition was established in 2006 by Sandra Latusseck and Soenke Schulz. Based in Hamburg, Germany, tredition offers publishing solutions to authors and publishing houses, combined with worldwide distribution of printed and digital book content. tredition is uniquely positioned to enable authors and publishing houses to create books on their own terms and without conventional manufacturing risks.

For more information please visit: www.tredition.com

TREDITION CLASSICS

This book is part of the TREDITION CLASSICS series. The creators of this series are united by passion for literature and driven by the intention of making all public domain books available in printed format again - worldwide. Most TREDITION CLASSICS titles have been out of print and off the bookstore shelves for decades. At tredition we believe that a great book never goes out of style and that its value is eternal. Several mostly non-profit literature projects provide content to tredition. To support their good work, tredition donates a portion of the proceeds from each sold copy. As a reader of a TREDITION CLASSICS book, you support our mission to save many of the amazing works of world literature from oblivion. See all available books at www.tredition.com.

Project Gutenberg

The content for this book has been graciously provided by Project Gutenberg. Project Gutenberg is a non-profit organization founded by Michael Hart in 1971 at the University of Illinois. The mission of Project Gutenberg is simple: To encourage the creation and distribution of eBooks. Project Gutenberg is the first and largest collection of public domain eBooks.

Sir Jagadis Chunder Bose His Life and Speeches

Jagadis Chandra, Sir Bose

Imprint

This book is part of TREDITION CLASSICS

Author: Jagadis Chandra, Sir Bose
Cover design: Buchgut, Berlin – Germany

Publisher: tredition GmbH, Hamburg - Germany
ISBN: 978-3-8472-3378-7

www.tredition.com
www.tredition.de

Copyright:
The content of this book is sourced from the public domain.

The intention of the TREDITION CLASSICS series is to make world literature in the public domain available in printed format. Literary enthusiasts and organizations, such as Project Gutenberg, worldwide have scanned and digitally edited the original texts. tredition has subsequently formatted and redesigned the content into a modern reading layout. Therefore, we cannot guarantee the exact reproduction of the original format of a particular historic edition. Please also note that no modifications have been made to the spelling, therefore it may differ from the orthography used today.

CONTENTS

His Life and Career
Literature and Science
Marvels of Plant Life
Plant Autographs — How Plants can record their own story
Invisible Light
Lecture on Electric Radiation
Plant Response
Evidence before the Public Services Commission
Prof. J. C. Bose at Madura
Prof. J. C. Bose Entertained — Party at Ram Mohan Library
History of a Discovery
A Social Gathering
Light Visible and Invisible
Hindu University Address
The History of a Failure that was Great
Quest of Truth and Duty
The Voice of Life
The Praying Palm of Faridpur
Visualisation of Growth
Sir J. C. Bose at Bombay
Unity of Life
The Automatic Writing of the Plant
Control of Nervous Impulse
Marvels of Growth as Revealed by the "Magnetic Crescograph"
The Night-Watch of Nymphaea
Wounded Plants

SIR JAGADIS CHUNDER BOSE

On the 30th November, 1858, Jagadis Chunder was born, in a respectable Hindu family, which hails from village Rarikhal, situated in the Vikrampur Pargana of the Dacca District, in Bengal. He passed his boyhood at Faridpur, where his father, the late Babu Bhugwan Chunder Bose, a member of the *then* Subordinate Executive Service was the Sub-Divisional Officer; and it was there that he derived "the power and strength that nerved him to meet the shocks of life." [1]

HIS FATHER

His father was a fine product of the Western Education in our country. Speaking of him, says Sir Jagadis "My father was one of the earliest to receive the impetus characteristic of the modern epoch as derived from the West. And in his case it came to pass that the stimulus evoked the latent potentialities of his race for evolving modes of expression demanded by the period of transition in which he was placed. They found expression in great constructive work, in the restoration of [Pg_002] quiet amidst disorder, in the earliest effort to spread education both among men and women, in questions of social welfare, in industrial efforts, in the establishment of people's bank and in the foundation of industrial and technical schools." [2] However, his efforts—like most pioneer efforts—failed. He became overpowered in the struggle. But his young son, who witnessed the struggle, derived a great lesson which enabled him "to look on success or failure as one"—or rather "failure as the antecedent power which lies dormant for the long subsequent dynamic expression in what we call success." "And if my life" says Sir Jagadis "in any way came to be fruitful, then that came through the realisation of this lesson." [2] So great was the influence exerted on him by his father that Sir Jagadis Chunder has observed "To me his life had been one of blessing and daily thanksgiving." [2]

HIS EARLY EDUCATION

Little Jagadis received his first lesson in a village *pathsala*. His father, who had very advanced views in educational matters, instead of sending him to an English School, which was then regarded as the only place for efficient instruction, sent him to the vernacular village school for his early education. [Pg_003] "While my father's subordinates" says Sir Jagadis "sent their children to the English schools intended for gentle folks, I was sent to the vernacular school, where my comrades were hardy sons of toilers and of others who, it is now fashion to regard, were belonging to the depressed classes." [3] Speaking of the effect it produced on him, observes Sir Jagadis "From these who tilled the ground and made the land blossom with green verdure and ripening corn, and the sons of the fisher folk, who told stories of the strange creatures that frequented unknown depths of mighty rivers and stagnant pools, I first derived the lesson of that which constitutes true manhood. From them too I drew my love of nature." [3]

"I now realise" continues Sir Jagadis "the object of my being sent at the most plastic period of my life to the vernacular school where I was to learn my own thoughts and to receive the heritage of our national culture through the medium of our own literature. I was thus to consider myself one with the people and never to place myself in an equivocal position of assumed superiority." [3]

"The moral education which we received in our childhood" adds Sir Jagadis "was very indirect and came from listening to stories recited by the [Pg_004] "Kathaks" on various incidents connected with our great epics. Their effects on our mind was Very great." [4]

And it is very interesting to learn from the lips of Sir Jagadis himself "that the inventive bent of his mind received its first impetus" in the industrial and technical schools established by his father. [4]

HIS COLLEGIATE EDUCATION IN INDIA

After he had developed, in the *pathsala*, some power of observation, some power of reasoning and some power of expression through the healthy medium of his own mother tongue, young Jagadis was sent to an English School for education. He passed the

Entrance Examination, in 1875, from the St. Xavier's Collegiate School, Calcutta, in the First Division. He then joined the College classes of that Institution, and there, in the "splendid museum of Physical Science Instruments," he drew his early inspirations in Physics from that remarkable educationist and brilliant experimentalist, the Rev. Father E. Lefont, S.J., C.I.E., M.I.E.E., who had the rare gift of enkindling the imagination of his pupils. He passed the First Examination in Arts, in 1877, in the Second Division and the B.A. Examination by the B. Course (Science Course), in 1880, in the [Pg_005] Second Division. "It is the paramount duty of the University" says Sir Ashutosh Mookerjea "to discover and develop unusual talent." [5] The Calcutta University, by the test of examination which it applied, totally failed to *discover* (not to speak of *developing*) the powers of an original mind which was destined to enrich the world by giving away the fruits of its experience.

HIS STUDY ABROAD

After Jagadis had graduated himself, in the Calcutta University, he longed to get a course of scientific education in England. He was sent to Cambridge and joined the Christ's College. He came in "personal contact with eminent men, whose influence extorted his admiration and created in him a feeling of emulation. In the way he owed a great deal to Lord Rayleigh, under whom he worked." [6] He passed the B.A. Examination of the Cambridge University, in Natural Science Tripos, in 1884. He also secured, in 1883, the B.Sc. Degree with Honours of London University. Jagadis had, by birth, the speculative Indian mind. And, by his scientific education, at home and abroad, he developed a capacity for accurate experiment [Pg_006] and observation and learnt to control his Imagination—"that wonderous faculty which, left to ramble uncontrolled leads us astray into a wilderness of perplexities and errors, a land of mists and shadows; but which, properly controlled by experience and reflection, becomes the noblest attribute of man; the source of poetic genius, the instrument of discovery in Science." [7] His strength and fertility as a discoverer is to be referred in a great measure to the harmonious blending of the burning Imagination of the East with the analytical methods of the West.

APPOINTED AS A PROFESSOR

After having completed his education abroad. Jagadis chose the teaching of Science as his vocation. He was appointed as Professor of Physical Science at the Presidency College, Calcutta. He joined the service on the 7th January, 1885. Although he was appointed in Class IV of the *then* Bengal Educational Service, (which afterwards merged in the present Indian Educational Service), he was not admitted to the full scale of pay of the Service. He, being an Indian, was allowed to draw only two-thirds the pay of his grade. This humiliating distinction was, however, removed in [Pg_007] his case, on the 21st September 1903, when the bureaucracy could not any longer ignore the pressure of enlightened opinion that was brought to bear on it.

HIS RESEARCHES ON ELECTRIC WAVES

It was in 1887, some times after Professor J. C. Bose had joined the Presidency College, Hertz demonstrated, by direct experiment, the existence of Electric Waves—the properties of which had been predicted by Clerk Maxwell long before. This great discovery sent a reverberation through the gallery of the scientific world. And, at once, the scientists in all countries began to devote their best energies to explorations in this new Realm of Nature. Young J. C. Bose— who had drunk deep at the springs of Scientific Knowledge and whose imagination had been very deeply touched by the scientific activities of the West and who had in him the burning desire that India should 'enter the world movement for that advancement of knowledge'—also followed suit.

DIFFICULTIES OF RESEARCHES

When, however, Prof. J. C. Bose joined the Presidency College, there was no laboratory worth the name there, nor had he any of 'those mechanical facilities at his disposal which every prominent European and American experimental scientist [Pg_008] commands'. He had to work under discouraging difficulties before he could begin his investigations. He was, however, not a man to quarrel with circumstances. He bravely accepted them and began to work in his own private laboratory and with appliances which, in any other country, would be deemed inadequate. He applied himself closely to the investigation of the invisible etheric waves and, with the simple means at his command, accomplished things, which few were able to perform in spite of their great wealth of external appliances.

As the wave-length of a Hertzian (electric) ray was very large — about 3 metres [8] long — compared with that of visible light, considerable difficulties were experienced in carrying on experiments with the same. It was thought, for instance, that very large crystals, much larger than what occur in nature, would be required to show the polarisation of electric ray. Prof. Bose who 'combined in him the inventiveness of a resourceful engineer, with the penetration and imagination of a great scientist' — designed an instrument which generated very short electric waves with a length of about 6 millimetres or so. And, by working with Electric radiations having very short wave-lengths, he succeeded in demonstrating that the electric waves are polarised by the crystal *Nemalite* (which he [Pg_009] himself discovered) in the very same way as a beam of light is polarised by the crystal Tourmaline. He then showed that a large number of substances, which are opaque to Light (*e.g.* pitch, coal-tar etc.) are transparent to Electric Waves. He next determined the Index of Refraction of various substances for invisible Electric Radiation and thereby eliminated a great difficulty which had presented itself in Maxwell's theory as to the relation between the index of refraction of light and the di-electric constant of insulators. He then determined the wave length of Electric Radiation as produced by various oscillators.

HIS EARLY CONTRIBUTIONS AND THEIR APPRECIATIONS

His first contribution was 'On Polarisation of Electric Rays by Double Refracting Crystals.' It was read at a meeting of the Asiatic Society of Bengal, held on the 1st May 1895, and was published in the Journal of the Society in Vol. LXIV, Part II, page 291. His next contributions were 'On a new Electro polariscope' and 'On the Double Refraction of the Electric Ray by a Strained Di-electric.' They appeared, in the *Electrician*, the leading journal on Electricity, published in London. These 'strikingly original researches' won the [Pg_010] attention of the scientific world. Lord Kelvin, the greatest physicist of the age, declared himself 'literally filled with wonder and admiration for so much success in the novel and difficult problem which he had attacked.' Lord Rayleigh communicated the results of his remarkable researches to the Royal Society. And the Royal Society showed its appreciation of the high scientific value of his investigation, not only, by the publication, with high tributes, of a paper of his 'On the Determination of the Indices of Electric Refraction,' in December 1896, and another paper on the 'Determination of the Wave-length of Electric Radiation,' in June 1896, but also, by the offer, of their own accord, of an appropriation from the Special Parliamentary Grant made to the Society for the Advancement of Knowledge, for continuation of his work.

In recognition of the importance of the contribution made by Prof. Bose, the University of London conferred on him the Degree of Doctor of Science and the Cambridge University, the degree of M.A., in 1896. And, to crown all, the Royal Institution of Great Britain—rendered famous by the labour of Davy and Faraday, of Rayleigh and Dewar—honoured him by inviting to deliver a 'Friday Evening Discourse' on his original work. It would not be out of place to observe that the rare privilege of being invited to deliver a 'Friday Evening [Pg_011] Discourse' is regarded as one of the highest distinction that can be conferred on a scientific man.

HIS FIRST SCIENTIFIC DEPUTATION. (1896-97)

The Government of India showed its appreciation of his work by deputing him to Europe to place the results of his investigations before the learned Scientific Bodies. He remained on his Deputation from the 22nd July 1896 to the 19th April 1897. He read a paper 'On a complete Apparatus for studying the Properties of Electric Waves' at the meeting of British Association, held at Liverpool, in 1896. He then communicated a paper 'On the Selective Conductivity exhibited by Polarising Substances,' which was published by the Royal Society, in January 1897. He next delivered his 'Friday Evening Discourse,' at the Royal Institution, 'On Electric Waves,' on the 29th January 1897. "There is, however, to our thinking" wrote the *Spectator* at the time "something of rare interest in the spectacle presented of a Bengalee of the purest descent possible, lecturing in London to an audience of appreciative European savants upon one of the most recondite branches of the modern physical science." He was then invited to address the Scientific Societies in Paris. "Prof. J. C. Bose" wrote the Review Encyclopedique, Paris "exhibited on the 9th of March before the Sorbonne, [Pg_012] an apparatus of his invention for demonstrating the laws of reflection, refraction, and polarisation of electric waves. He repeated his experiments on the 22nd, before a large number of members of the Academie des Sciences, among whom were Poincare, Cornu, Mascart, Lipmann, Cailletet, Becquerel and others. These savants highly applauded the investigations of the Indian Professor." M. Cornu, President of the Academy of Science, was pleased to address Professor Bose as follows:—

"By your discoveries you have greatly furthered the cause of Science. You must try to revive the grand traditions of your race which bore aloft the torch light of art and science and was the leader of civilization two thousand years ago. We, in France applaud you." This fervent appeal, we shall see, as we proceed, did not go in vain.

He was next invited to lecture before the Universities in Germany. At Berlin, before the leading physicists of Germany, he gave an address on Electric Radiation, which was subsequently published in the *Physikaliscen Gesellschaft Berlin*, in April 1897.

FURTHER RESEARCHES ON ELECTRIC WAVES

Having received the most generous and wide appreciation of his work, Dr. J. C. Bose continued, with redoubled vigour, his valuable researches on Electric Waves. He studied the influence of thickness [Pg_013] of air-space on total reflection of Electric Radiation and showed that the critical thickness of air-space is determined by the refracting power of the prism and by the wave-length of the electric oscillations. He next demonstrated the rotation of the plane of polarisation of Electric Waves by means of pieces of twisted jute rope. He showed that, if the pieces are arranged so that their twists are all in one direction and placed in the path of radiation, they rotate the plane of polarisation in a direction depending upon the direction of twists; but, if they are mixed so that there are as many twisted in one direction as the other, there is no rotation. [9] He communicated to the Royal Society the results of his new researches. And the Royal Society published, in November 1897, his papers 'On the Determination of the Index of Refraction of glass for the Electric Ray' and 'On the influence of Thickness of Air-space on Total Reflection of Electric Radiation' and, in March 1898, his further contributions 'On the Rotation of Plane of Polarisation of Electric Waves by a twisted structure' and 'On the Production of a "Dark cross" in the Field of Electro-magnetic Radiation.'

SELF-RECOVERING "COHERER"

The study of Electric Waves by Dr. J. C. Bose led [Pg_014] not only to the devising of methods for the production of the shortest Electric Waves known but also to the construction of a very delicate 'Receiver' for the detection of invisible other disturbances. The most sensitive form of detector hitherto known was the "Coherer." One of the forms made by Sir Oliver Lodge consisted simply of a glass tube containing iron turnings, in contact with which were wire led into opposite ends of the tube. The arrangement was placed in series with a galvanometer and a battery; when the turnings were struck by electric waves, the resistance between loose metallic contacts was diminished and the deflection of the galvanometer was increased. Thus the deflection of the galvanometer was made to indicate the arrival of electric waves. The arrangement was, no doubt, a sensitive one, but, to get a greater delicacy, Dr. Bose used, instead of iron turnings, spiral springs which were pushed against each other by means of a screw. [10] Still the arrangement laboured under one great disadvantage. The 'receiver' had to be tapped between each experiment. So something better than a 'cohering' receiving was needed — something that was self-recovering, like a human eye. To discover that something, Dr. Bose began a study of the whole theory of 'coherer action.' It was hitherto believed [Pg_015] that the electric waves, by impinging on iron and other metallic particles in contact, brought about a sort of fusion — a sort of 'coherence' — and that the diminution of resistance was the result of that 'coherence.' To satisfy himself as to the correctness of this theory, Dr. Bose engaged himself in a most laborious investigation to find out the action of electric radiation not only on iron particles but on all kinds of matter and ultimately discovered the surprising fact that, though the impact of electric waves generally produced a diminution of resistance, with *potassium* there was an *increase* of resistance after the waves had ceased. [11] This discovery at once showed the untenability of the old theory and pointed to the conclusion that the effect of electric radiation on matter is one of discriminative molecular action—that the Electric Waves produced a re-arrangement of the molecules which may either increase or decrease the contact resistance. It may be incidentally mentioned here that this detection of molecular change in matter under electric stimulation has given rise to a new theory of photographic action.

As a result of his painstaking investigation on the action of Electric Waves on different kinds of matter, Dr. Bose invented a new type of self-recovering electric receiver, "so perfect in its action [Pg_016] that the Electrician suggested its use in ships and in electro-magnetic light-houses for the communication and transmission of danger-signals at sea through space. This was, in 1895, several years in advance of the present wireless system." Practical application of the results of Dr. Bose's investigations appeared so important that the Governments of Great Britain and the United States of America granted him patents for his invention of a certain crystal receiver which proved to be the most sensitive detector of the wireless signal. Dr. Bose, however, has made no secret at any time as to the construction of his apparatus. He has never utilised the patents granted to him for personal gain. His inventions are "open to all the world to adopt for practical and money-making purposes." "The spirit of our national culture" observes Sir J. C. Bose "demands that we should for ever be free from the desecration of utilising knowledge for personal gain." [12]

HIS RESEARCHES TAKE A NEW TURN

This inquiry which Dr. J. C. Bose started for the purpose of ascertaining 'coherer action'—why the "receiver" had to be tapped in order to respond again to electric waves — took him [Pg_017] unconsciously to the border region of physics and physiology and gave an altogether new turn to his researches. "He found that the uncertainty of the early type of his receiver was brought on by 'fatigue' and that the curve of fatigue of his instrument closely resembled the fatigue curve of animal muscle." [13] He did not stop there but pushed on his investigations and found "that the 'tiredness' of his instrument was removed by suitable stimulants and that application of certain poisons, on the other hand, permanently abolished its sensitiveness." He was amazed at this discovery — this parallelism in the behaviour of the 'receiver' to the living muscle. This led him to a systematic study of all matter, Organic and Inorganic, Living and Non-Living.

RESPONSE IN LIVING AND NON-LIVING

He began an examination of inorganic matter in the same way as a biologist examines a muscle or a nerve. He subjected metals to various kinds of stimulus — mechanical, thermal, chemical, and electrical. He found that all sorts of stimulus produce an excitatory change in them. And this excitation sometimes expresses itself in a visible change of [Pg_018] form and sometimes not; but the disturbance produced by the stimulus always exhibits itself in an *electric response*. He next subjected plants and animal tissues to various kinds of stimulus and also found that they also give an *electric response*. Finding that a universal reaction brought together metals, plants and animals under a common law, he next proceeded to a study of *modifications in response*, which occur under various conditions. He found that they are all benumbed by cold, intoxicated by alcohol, wearied by excessive work, stupified by anaesthetics, excited by electric currents, stung by physical blows and killed by poison — they all exhibit essentially the same phenomena of fatigue and depression, together with possibilities of recovery and of exaltation, yet also that of permanent irresponsiveness which is associated with death — they all are responsive or irresponsive under the same conditions and in the same manner. The investigations showed that, in the entire range of response phenomena (inclusive as that is of metals, plants and animals) there is no breach of continuity; that "the living response in all its diverse modifications is only a repetition of responses seen in the inorganic" and that the phenomena of response "are determined, not by the play of an unknowable and arbitrary *vital force*, but by the working of laws that know no change, acting [Pg_019] equally and uniformly throughout the organic and inorganic matter." [14]

SECOND SCIENTIFIC DEPUTATION, 1900-01

In the year 1900, the International Scientific Congress was held, in Paris. And Dr. J. C. Bose was deputed by the Government of India to the Congress as a delegate from this country. Before the assembled scientists, Dr. Bose delivered a remarkable address on the results of his researches on the similarity of Response of Inorganic and Living Substances to Electric stimulus ... 'De la gjnjralitj de Phjnomjnes Moleculairs produits par l'Ectriciti sur la matirij Inorganique et sur la matijre Vivante.' He next read a paper 'On the Similarity of effect of Electric Stimulus on Inorganic and Living Substances' before the Bradford meeting of the British Association in 1900. He then contributed a very interesting paper 'on Binocular Alteration of Vision,' which was published by the Physiological Society of London, in November 1900. It may be mentioned here, by the way, that, in course of his investigations on the Response of the Living and Non-Living substances, Dr. Bose constructed an "artificial retina" to study the characteristics of the excitatory change produced by a stimulus on [Pg_020] the retina and these characteristics gave him a clue to the unexpected discovery of the "binocular alteration of vision" in man—"each eye supplements its fellow by turns, instead of acting as a continuously yoked pair, as hitherto believed." [15] He next communicated to the Royal Society his researches 'On the Continuity of Effect of Light and Electric Radiation on Matter,' and 'On the Similarities between Mechanical and Radiation Strains,' and 'On the Strain Theory of Photographic action,' which were published in April 1901. Then, on the 10th May 1901, he delivered his remarkable 'Friday Evening Discourse,' at the Royal Institution, on the 'Response of Inorganic Matter to Stimulus.'

OPPOSITION OF THE PHYSIOLOGISTS

Then, on the 5th June 1901, he gave an experimental demonstration, before the Royal Society, on the subject of his researches 'On Electric Response of Inorganic Substances' which had already been communicated to that Society, on the 7th May 1901. He was strongly assailed by Sir John Burden Sanderson, the leading physiologist, and some of his followers. They objected to a physicist straying into the preserve especially reserved for [Pg_021] them. They dogmatically asserted *as physiologists* that the excitatory response of ordinary plants to mechanical stimulus was an impossibility. But they failed to urge anything against the experiment of the physicist. In consequence of this opposition, Dr. Bose's paper, which was already in print, was not published but was placed in the archives of the Royal Society. "And it happened that eight months after the reading of his Paper, another communication found publication in the Journal of a different Society which was practically the same as Dr. Bose's but without any acknowledgment. The author of this communication was a gentleman who had previously opposed him at the Royal Society. The plagiarism was subsequently discovered and led to much unpleasantness. It is not necessary to refer any more to this subject except as an explanation of the fact that the determined hostility and misrepresentation of one man succeeded for more than 10 years to bar all avenues of publications for his discoveries." [16]

The opposition of the physiologists, however, did one good. It spurred Dr. Bose on and made him stronger in his determination not to encompass himself, within the narrow groove of physical investigation. He took furlough for one year, in [Pg_022] extension of the period of his Deputation, and applied himself vigorously to the investigations, which he had already commenced in India and received facilities from the Managers of the Royal Institution to work in the Davy-Faraday Laboratory. He next read, at the Glasgow meeting of the British Association, in 1901, a paper 'On the Conductivity of Metallic particles under Cyclic Electro-magnetic Variation.' Then, in March 1902, "Prof. Bose" says the *Nature* "performed a series of experiments before the Linnean Society showing electric response for certain portions of the plant organism, which proved that as concerning fatigue, behaviour at high and low temperatures, the effects produced by poisons and anaesthetics, the responses are identical with those held to be characteristic of muscle and nerve." The Linnean Society published, in its Journal, in March 1902, his paper 'On Electric Response of Ordinary Plants under Mechanical Stimulus.' He then communicated to the Sociiti de Physique, Paris, his paper 'Sur la Rispouse Electrique dans les Mitaux, les Tissu Animaux et Vigitaux.' The Royal Society published, in April 1902, his contribution 'On the Electromotive Wave accompanying Mechanical Disturbance in Metals in contact with Electrolyte.' He was next asked by the Royal Photographic Society to give a [Pg_023] discourse 'On the Strain Theory Vision and of Photographic Action,' which was published by the Society, in its Journal, in June 1902. He then wrote a paper 'On the Electric Response in Animal, Vegetable and Metal,' which was read before the Belfast meeting of the British Association, in 1902. The President of the Botanical Section at Belfast, in his address, observed "Some very striking results were published by Bose on Electric Response in ordinary plants. Bose's investigations established a very close similarity in behaviour between the vegetable and the animal. Summation effects were observed and fatigue effect demonstrated, while it was definitely shown that the responses were physiological. They ceased as soon as the piece of tissue was killed by heating. These observations strengthen considerably the view of the identical nature of the animal and vegetable protoplasm."

Dr. Bose then brought out a systematic treatise embodying the results of his researches under the significant title of 'Response in the Living and Non-living.' He returned to India, in October, 1902.

GOVERNMENT RECOGNITION

After he had come back, from the Second Scientific Deputation, the Government of India conferred on him the distinction of Companion of the Order of [Pg_024] the Indian Empire, in 1903, in recognition of his valuable researches.

PLANT LIFE AND ANIMAL LIFE

Next Dr. Bose, in natural sequence to the investigation of the response in 'inorganic' matter commenced 'a prolonged study of the activities of plant life as compared with corresponding functioning of animal life.'

ALL PLANTS ARE "SENSITIVE"

It was believed that so-called 'sensitive' plants alone exhibited excitation by *electric response*. But Dr. Bose, believing in continuity of responsive phenomena, used the same experimental devices, with which he had already succeeded in obtaining the *electric response* of inorganic substances, to test whether ordinary plants also—meaning those usually regarded as 'insensitive'—would or would not exhibit excitatory *electrical response* to stimulus. With the help of very delicate instruments, Dr. Bose demonstrated the very startling fact that not only every plant, but every organ of every plant gave true *excitatory electric response*—and that response was not confined alone to 'sensitive' plants like *Mimosa*.

Dr. Bose then proceeded to investigate whether the responsive effects which he had shown to occur [Pg_025] in ordinary plants might not be further exhibited by means of *visible mechanical response*, thus fully removing the distinction commonly assumed to exist between the 'sensitive' and supposed 'non-sensitive.' Dr. Bose invented 'special apparatus of extreme delicacy,' which detected infinitesimal tremors, and showed that ordinary plants, usually regarded as insensitive, gave *motile responses*, which had hitherto passed unnoticed. His later investigation shows that "all plants, even the trees, are fully alive to changes of environment; they respond visibly to all stimuli, even to the slight fluctuations of light by a drifting cloud." [17]

'TROPIC' MOVEMENTS

Finding that the plants give, not only *electric* but *motile* response as well, to stimulus, Dr. Bose proceeded to study the nature of responses evoked in plants by the *stimuli of the natural forces*. He found that plants respond visibly, by movements, to *environmental stimuli*. But the movements induced—'tropic' movements—are extremely diverse. Light, for example, induces sometimes positive curvature, sometimes negative. Gravitation, again, induces one movement in the root, and the opposition in the shoot. Dr. Bose applied himself to find out whether [Pg_026] the movements in response to external stimuli, though apparently so diverse, could not be ultimately reduced to a fundamental unity of reaction. As a result of a very deep and penetrating study of the effects of various environmental stimuli, on different plant organs, he showed that the cells on two sides are unequally influenced, on account of different external conditions, and contract unequally, and hence the various movements are produced—that the many anomalous effects, hitherto ascribed to 'specific sensibilities,' are due to the 'differential sensibilities'—differential excitability of anisotropic structures and to the opposite effects of external and internal stimuli—that all varieties of plant movements are capable of a consistent mechanical explanation. Dr. Bose's "latest investigations recently communicated to the Royal Society have established the single fundamental reaction which underlies all these effects so extremely diverse." [18]

EXTENDED APPLICATION OF MECHANICAL THEORY

With an extended application of his mechanical theory, Dr. Bose has gradually removed the veil of obscurity from many a phenomenon in plant life. [Pg_027] The 'autonomous' movements of plants, for example, which remained enveloped in mystery, received a satisfactory solution at his hands.

'AUTONOMOUS' MOVEMENTS

It was believed that automatically pulsating tissues draw their energy from a mysterious "vital force" working within. By controlling external forces, Dr. Bose stopped the pulsation and re-started it and thus demonstrated that the 'automatic action' was not due to any internal vital force. He pointed out that the external stimulus—instead of causing, as was customary to suppose, an explosive chemical change and an inevitable run-down of energy—brings about an accumulation of energy by the plant. And with the accumulation of absorbed energy, a point is reached when there is an overflow — the excess of energy bubbles over, as it were, and shows itself in 'spontaneous' movements. The stimulus being strong a single response — a single twitching of the leaflets — is not enough to express the whole of the leaf's responsive energy and it yields a multiple response — it reverberates — it manifests itself in 'automatic' pulsations. When, however, the accumulated energy is exhausted, then there is also an end of 'spontaneous movements.' There are strictly speaking, no 'spontaneous' movements; those known by that name are really due [Pg_028] either to the immediate effects of external stimulus or to the stimulus previously absorbed and held latent in the plant to find subsequent expression — due to the direct or indirect action of external forces which are transformed in the machinery of the plants in obedience to the principle of the Conservation of Energy.

"ASCENT OF SAP" "AND GROWTH"

Dr. Bose then showed that, not gross mechanical movements alone, but also other invisible movements are initiated by the action of stimulus, and that the various activities, such as the "ascent of sap" and "growth" are in reality different reactions to the stimulating action of energy supplied by the environment. In this way, Dr. Bose showed that several obscure phenomena, in the life-processes of the plant, can be very satisfactorily explained by the Mechanical Theory.

It would not be out of place to mention that Dr. Bose, to carry on his researches on the Ascent of Sap, invented a new type of instrument (Shoshungraph). And for an accurate investigation on the phenomenon of growth of plants he devised an instrument (Growth Recorder) for instantaneous measurement of the rate of growth and another instrument (Balanced Crescograph) for determining the influences of various agencies on growth. So [Pg_029] very marvellous these instruments that the growth, which takes place, during a few beats of pendulum, is measured, and, in less than a quarter of an hour, the action of fertilizers, foods, electrical currents and various stimulants are determined. "What is the tale of Aladdin and his wonderful lamp" exclaims the Editor of the *Scientific American* "compared with the true story told by the crescograph?... Instead of waiting a whole season, perhaps years, to discover whether or not it is wise to mix this or that fertilizer with the soil one can now find in a few minutes!" Yet these are the instruments which are better known in Washington than in Calcutta! The question of their application to practical agriculture has excited more interest in the United States of America than in this unfortunate land, which is an essentially agricultural country!

FUNDAMENTAL IDENTITY OF REACTIONS

Dr. Bose showed that there is no physiological response given by the most highly organised animal tissue that is not also to be met with in the plant. He carried on "Researches on Diurnal Sleep" and showed that the plant is not equally sensitive to an external stimulus during day and night, and that there is a fundamental identity of life-reaction in plant and animal, as seen in a similar periodic [Pg_030] insensibility in both, corresponding to what we call *sleep*. He also showed that the passage of life in the plant, as in the animal, is marked by an unmistakable spasm. He invented, an instrument (Morograph) with which he recorded the critical point of death of a plant with great exactness. He demonstrated, in the most conclusive manner, that there is an essential unity of physiological effects of drugs on plant and animal tissues and showed the modifications which are introduced into these effects by the factor of individual 'constitution.' It may be mentioned casually that "this physiological identity in the effect of drugs is regarded by leading physicians as of great significance in the scientific advance of Medicine; since we have a means of testing the effect of drugs under conditions far simpler than those presented by the patient, far subtler too, as well as more humane than those of experiments on animals." [19] Dr. Bose further demonstrated that there is conduction of the excitatory impulse in the plant, like the nervous impulse in the animal; and showed the possibility of detecting the wave in transit and measured the speed with which the excitation coursed through the plant and also showed that the velocity of excitation is modified, by different agencies, even in the case [Pg_031] of ordinary plants. He also showed that the polar effects induced by electric currents, both in plants and animals, are identical.

These remarkable researches on Plant Response have 'revolutionised in some respects and very much extended in others our knowledge of the response of plants to stimulus.'

FURTHER DIFFICULTIES

Dr. Bose communicated his paper 'On the Electric Pulsation accompanying Automatic Movements in Desmodium Gyrans' to the Linnaean Society, which was published, in December 1902. Then, in 1903, he communicated to the Royal Society his researches on 'Investigation on Mechanical Response in Plants,' 'On Polar effects of Currents on the Stimulation of Plants,' 'On the Velocity of Transmission of Excitatory waves in Plants,' 'On the excitability and conductivity of Plant Tissues,' 'On the Propagation of the Electromotive Wave concomitant of Excitatory Waves in Plants,' 'On Multiple Response in Plants,' 'On an enquiry into the cause of Automatic Movements.'

"These new contributions" made by Dr. Bose on Plant Response "were regarded as of such great importance that the Royal Society showed its special appreciation by recommending them to be [Pg_032] published in their Philosophical Transactions. But the same influence, which had hitherto stood in his way, triumphed once more, and it was at the very last moment that the publication was withheld. The Royal Society, however, informed him that his results were of fundamental importance, but as they were so wholly unexpected and so opposed to the existing theories, that they would reserve their judgment until, at some future time, plants themselves could be made to record their answers to questions put to them. This was interpreted in certain quarters here as the final rejection of Dr. Bose's theories by the Royal Society and the limited facilities which he had in the prosecution of his researches were in danger of being withdrawn." [20]

HE BUILT HIS LIFE ON THE ROCK OF FAITH

But these difficulties—sufficient to crush many a spirit—could hardly quench the ardour of his burning soul, which was 'hungering and thirsting' for the establishment of a truth in which he had a firm Faith. Though the surges would beat against him, he would not give way. With the true spirit of a *Sadhak*, he devoted himself to the realisation of the great dream of his life. And, for the next ten years, the one *tap, jap* and *aradhana* of his life—the [Pg_033] one all-engrossing idea of his mind—was how to make the plant give testimony by means of its own autograph.

PUBLICATION OF "PLANT RESPONSE"

Though his researches did not find an outlet, in the Proceedings of the Royal Society, he did not lose heart. He brought out, in April 1906, a systematic treatise—"The Plant Response as a Means of Physiological Investigation"—in which he incorporated the results of his investigations on plant life.

ADOPTS A NEW METHOD OF INVESTIGATION

Hitherto Dr. Bose detected the various excitatory effects of plants by means of *mechanical response*. Being now confronted with opposition, he turned his attention to the finding of corroboration of the various results, which he had already obtained, by some other method of investigation. And for this he employed the method of *electric response*. He found that the results obtained by this new method of inquiry corroborated those already obtained by him by the old method. Emboldened by this corroboration, he next proceeded to extend this new method of inquiry by means of *electric response* into the field of Animal Physiology with a view to explain responsive phenomena in general on the [Pg_034] consideration of that fundamental molecular reaction which occurs even in inorganic matter.' [21]

RESULT OF THE INVESTIGATION

Dr. Bose found, in the plant as well as in the animal, "a similar series of excitatory effects, whether these be exhibited mechanically or electrically. Both alike are responsive, and similarly responsive, to all the diverse forms of stimulus that impinge upon them. We ascend, in the one case as in the other, from the simplicities of the isotropic to the complexities of the anisotropic; and the laws of these isotropic and anisotropic responses are the same in both. The responsive peculiarities of epidermis, epithelium, and gland; the response of the digestive organ, with its phasic alterations; and the excitatory electrical discharge of an anisotropic plate, are the same in the plant as in the animal. The plant, like the animal, is a single organic whole, all its different parts being connected, and their activities co-ordinated, by the agency of those conducting strands which are known as nerves. As in the plant nerve, moreover, so also in the animal, stimulation gives rise to two distinct impulses, exhibiting themselves by two-fold mechanical and electrical indications of [Pg_035] opposite signs.... The dual qualities or tones known to us in sensation, further, are correspondent with those two different nervous impulses, of opposite signs, which are occasioned by stimulation. These two sensory responses—positive and negative, pleasure and pain—are found to be subject to the same modifications, under parallel conditions, as the positive and negative mechanical and electrical indications with which they are associated. And finally, perhaps, the most significant example for the effect of induced anisotropy lies in that differential impression made by stimulus on the sensory surfaces, which remains latent, and capable of revival, as the memory-image. In this demonstration of continuity, then, it has been found that the dividing frontiers between Physics, Physiology, and Psychology have disappeared." [22]

CLASH WITH CURRENT VIEWS

The results, which Dr. Bose obtained from actual experiments, clashed, however, with the theories in vogue. The reactions of different issues were hitherto regarded as *special differences*. As against this, a *continuity* is shown to exist between them. Thus, nerve was universally regarded as [Pg_036] typically *non-motile*; its responses were believed to be characteristically different from those of muscle. Dr. Bose, however, has shown that nerve is indisputably motile and that the characteristic variations in the response of nerve are, generally speaking, similar to those of the muscle.

It was customary to regard plants as devoid of the power to conduct true excitation. Dr. Bose had already shown that this view was incorrect. He now showed, by experiment, that the response of the *isolated* vegetal nerve is indistinguishable from that of animal nerve, throughout a large series of parallel variations of condition. So complete, indeed, is the similarity between the responses of plant and animal, found, of which this is one instance, that the discovery of a given responsive characteristic in one case proves a sure guide to its observation in the other, and the explanation of phenomenon, under the simpler conditions of the plant, is found fully sufficient for its elucidation under the more complex circumstances of the animal. Dr. Bose found 'differential excitability' is widely present as a factor in determining the character of special responses and showed that many anomalous conclusions, with regard to the response of certain animal tissues, had arisen from the failure to take account of the 'differential excitability' of anisotropic organs. Hitherto [Pg_037] Pfluger's Law of the polar effects of currents was supposed to rest on secure foundations. But Dr. Bose showed that Pfluger's Law was not of such universal application as was supposed. He demonstrated that, above and below a certain range of electromotive intensity, the polar effects of currents are precisely opposite to those enunciated by Pfluger.

SENSATION

It was supposed that nervous impulse, which, must necessarily form the basis of sensation, was beyond any conceivable power of visual scrutiny. But Dr. Bose showed that this impulse is actually attended by change of form, and is, therefore capable of direct observation. He also showed that the disturbance, instead of being single, is of two different kinds—*viz.*, one of expansion (positive) and the other of contraction (negative)—and that, when the stimulus is feeble, the positive is transmitted, and, when the stimulus is stronger, both positive and negative are transmitted, but the negative, however, being more intense, masks the positive. He identified the wave of expansion travelling along the nerve with the tendency to pleasure, and the wave of contraction, with the tendency to pain. It thus appears that all pain contains an element of [Pg_038] pleasure, and that pleasure, if carried too far becomes pain—that "the tone of our sensation is determined by the intensity of nervous excitation that reaches the central perceiving organ."

MEMORY IMAGE AND ITS REVIVAL

Dr. Bose next pointed out that there remains, for every response, a certain residual effect. A substance, which has responded to a given stimulus, retains, as an after-effect, a 'latent impression' of that stimulus and this 'latent impression' is capable of subsequent revival by bringing about the original condition of excitation. The impress made by the action of stimulus, though it remains latent and invisible, can be revived by the impact of a fresh excitatory impulse.

Experimenting with a metallic *leaf*, Dr. Bose demonstrated the revival of a latent impression under the action of diffused stimulus. The investigation by Dr. Bose on the after-effects of stimulus has thrown some light on the obscure phenomenon, of 'memory.' It appears that, when there is a mental revival of past experience, the diffuse impulse of the 'will' acts on the sensory surface, which contains the latent impression and re-awakens the image which appears to have faded out. Memory is concerned, thus, with the after-effect of an impression [Pg_039] induced by a stimulus. It differs from ordinary sensation in the fact that the stimulus which evokes the response, instead of being external and objective, is merely psychic and subjective.

Dr. Bose has, by experimental devises, shown the possibility of tracing 'memory-impression' backwards even in inorganic matter, such latent impression being capable of subsequent revival. An investigation of the after-effects of stimulus, on living tissues would open out the great problem of the influence of past events on our present condition.

DEATH-STRUGGLE AND MEMORY REVIVAL

There is a wide-spread belief that, in the case of a sudden death-struggle, as for example, when drowning, the memory, of the past comes in a flash. "Assuming the correctness of this," says Sir Jagadis "certain experimental results which I have obtained may be pertinent to the subject. The experiment consisted in finding whether the plant, near the point of death, gave any signal of the approaching crisis. I found that at this critical moment a sudden electrical spasm sweeps through every part of the organism. Such a strong and diffused stimulation—now involuntary—may be expected in a human subject to crowd into one [Pg_040] brief flash a panoramic succession, of all the memory images latent in the organism." [23]

"COMPARATIVE ELECTRO-PHYSIOLOGY"

Dr. Bose published the results of these new researches, in 1907, in another remarkable volume, which was styled 'The Comparative Electro-Physiology.'

THIRD SCIENTIFIC DEPUTATION, 1907-08

After the publication of 'The Comparative Electro-Physiology,' the Government of India again sent Dr. Bose on a Scientific Deputation. He went over to England and America and placed the results of his researches before the learned Scientific Bodies. He read a paper 'On Mechanical Response of Plants' at the Liverpool meeting of British Association, in 1907. He then read a paper on 'The Oscillating Recorder for Automatic Tracing of Plant Movements' before the New York Academy of Sciences, and, in December 1908, he gave an address on 'Mechanical and Electrical Response in Plants,' at the Annual Meeting of the American Association for the Advancement of Science, held at Baltimore, and, in January 1909, he [Pg_041] delivered a lecture on 'Growth Response of Plants' before the United States Department of Agriculture and, in February 1909, he read a paper on 'Death-spasm in Plants,' before the University of Illinois, and, in March 1909, a paper on 'Multiple and Autonomous Response in Plants' before the Madison University. He also lectured before the New York Botanical Society, the Medical Society of Boston, the Society of Western Electric Engineers at Chicago. He also delivered a series of post-graduate lectures on Electro-Physics and Plant Physiology at the Universities of Wisconsin, Chicago, Ann Arbor. He returned to India, in July 1909.

FURTHER EXPERIMENTAL EXPLORATION

By his new and newer methods of investigation, Dr. Bose got a deep and deeper perception of that underlying unity, for the demonstration of which he had been labouring since 1901. But the dream of his life was not yet realised. No direct method of obtaining response record was yet obtained. Hitherto the response recorder employed was a modification of the optical lever, automatic records being secured by the very inconvenient and tedious process of photography (which again introduced complications by subjecting a plant to darkness and [Pg_042] thereby modifying its normal excitability); and the plant was not automatically excited by stimulus, besides the results obtained were liable to be influenced by personal factor. So Dr. Bose set about the invention of an apparatus, which should discard the use of photography and in which the plant (attached to the recording apparatus) should be automatically excited by stimulus absolutely constant, should make its own responsive record, going through its own period of recovery, and embarking on the same cycle over again without assistance at any point on the part of the observer. Great difficulties were encountered in realising these ideal requirements. They appeared, at first, to be insurmountable. But, with continuous toil and persistence, Dr. Bose succeeded in designing a long battery of supersensitive instruments and apparatus, which made the seeming impossible possible. His ingenious "Resonant and Oscillating Recorders" gave a simple and direct method of obtaining the record. The plant, being automatically excited by stimulus, made its own responsive record. The closed doors, at last, opened. The secret of plant life stood revealed by the autographs of the plant itself. The great *sadhana* of his life now received its fulfilment. "It has been beautifully said—and it is a law of the moral world as unchangeable as physical laws—'Ask, and it shall be given you; [Pg_043] seek, and ye shall find; knock, and it

shall be opened unto you; for every one that asketh receiveth; and he that seeketh findeth and to him that knocketh it shall be opened."
24

TRANSMISSION OF EXCITATION IN MIMOSA

Dr. Bose had shown that all plants are sensitive—that there is no difference between the so-called 'sensitive' and the supposed 'non-sensitive'—that they gave alike the true excitatory *electric response* as well as *motile response*. The evidence of plant's script now removed beyond any doubt the long-standing error which divided the vegetable world into 'sensitive' and 'insensitive.' There remained, however, the question of nervous impulse in plants, the discovery of which, though announced by Dr. Bose, ten years ago, did not yet find full acceptance.

Finding that the scope of his investigation has been very much enlarged by the devise of the Resonant Recorder, Dr. Bose proceeded to attack the *current* view "that there was no transmission of true excitation in Mimosa, the propagated impulse being regarded as merely hydromechanical." This conclusion was based on the experiments of the leading German plant physiologists, Pfeffer and [Pg_044] Haverlandt who failed to bring on any variation in the propagated impulse in plants either by scalding or by application of an anaesthetic. Dr. Bose pointed out that, as Pfeffer applied the chloroform to the *outer* stalk and Haverlandt scalded the *outer* stem, neither the stimulant nor the anaesthetic reached the nerves. So he, instead of applying the stimulant or the anaesthetic, in the *liquid* form, to the outer stalk or stem, confined the Mimosa, in a little chamber, and subjected it to the influence of the *vapour* of the drug. The fumes now penetrated and reached the nerves and the plant was made to record, by its own script, the variations, if any, produced by the drugs. The plant, by its self-made records, showed exultation with alcohol, depression with chloroform, rapid transmission of a shock with the application of heat, and an abolition of the propagated impulse with the application of a deadly poison like potassium cyanide. This variation in the transmitted impulse, under physiological variations, showed that it was not a physical one. This sealed the fate of the hydromechanical theory.

Dr. Bose went further and showed that the impulse is transmitted in both directions along the nerve but not at the same rate. And, by interposing an electric block, he arrested the nervous impulse in a plant in a manner similar to the corresponding arrest in the animal nerve and thereby [Pg_045] produced nervous *paralysis* in plant, such paralysis being afterwards cured by appropriate treatment. "If he had made no other discovery," says the Editor of the *Scientific American* "Dr. Bose would have earned an enduring reputation in the annals of science. We know very little about paralysis in the human body, and practically nothing about its cause. The nervous system of the higher animals is so complicated, so intricate, that it is hard to understand its derangement. The human nerve dies when isolated. It is killed by the shock of removal, and responds for the moment abnormally and therefore deceptively. But, if we study the simplest kind of a nerve, — and the simplest is that of a plant, — we may hope to understand what occurs when a hand or a foot cannot be made to move. To find out that plants have nerves, to induce paralysis in such nerves and then to cure them — such experiments will lead to discoveries that may ultimately enable physicians to treat more rationally than they do, the various forms of paralysis now regarded as incurable."

MIMOSA AND MAN

Dr. Bose showed not only that the nervous impulse in plant and in man is exalted or inhibited under identical conditions but carried the parallelism [Pg_046] very far and pointed out the blighting effects on life of a complete seclusion and protection from the world outside. "A plant carefully protected under glass from outside shocks", says Sir Jagadis "looks sleek and flourishing; but its higher nervous function is then found to be atrophied. But when a succession of blows is rained on this effete and bloated specimen, the shocks themselves create nervous channels and arouse anew the deteriorated nature. And is it not shocks of adversity, and not cotton-wool protection, that evolve true manhood?" [25]

ROYAL SOCIETY

Having found that his investigation on Mimosa had broken down the barriers which separated kindred phenomena, Dr. Bose next communicated the results of his wonderful researches to the Royal Society. His paper was read, at a meeting of the Society, held on the 6th March 1913. The Royal Society *now* found that Dr. Bose had rendered the seemingly impossible, possible—had made the plant tell its own story by means of its self-made records. It could no longer withhold the recognition which was his due. The barred gates, at last, opened and the paper of Dr. Bose "On an Automatic Method, for the investigation of the Velocity of Transmission [Pg_047] of Excitation in Mimosa" found publication in the "Philosophical Transactions of the Royal Society" in Vol. 204, Series B.

HIS FURTHER INVESTIGATIONS

Dr. Bose next pursued with great vigour his investigations on the Irritability of Plants. By making the plant tell its own story, by means of its self-made records, he showed that there is hardly any phenomenon of irritability observed in the animal which is not also found in the plant and that the various manifestations of irritability in the plant are identical with those in the animal and that many difficult problems in Animal Physiology find their solution in the experimental study of corresponding problems under simpler conditions of vegetable life.

HOURS OF SLEEP OF THE PLANT

It may be mentioned that Dr. Bose showed one very remarkable fact—from the summaries of the automatic records of the responses given by a plant (which was subjected to an impulse during all hours of the day and night)—that it wakes up during morning slowly, becomes fully alert by noon, and becomes sleepy only after midnight, resembling man in a surprising manner.

"IRRITABILITY [Pg_048] OF PLANTS"

Dr. Bose embodied the results of his fascinating researches, obtained by the introduction of new methods, in another remarkable volume—"Researches on Irritability of plants"—which was published, in 1913.

FURTHER RECOGNITION

In recognition of his valuable researches, Dr. J. C. Bose was invested with the insignia of the Companion of the Order of the Star of India by His Majesty the King Emperor, on the occasion of his Coronation Durbar, at Delhi, in 1911.

The *intelligentsia* of Bengal showed also their tardy appreciation by calling on him to preside over the deliberations of the Mymensing meeting of the Bengal Literary Conference, held on the 14th April 1911, when he delivered a unique Address, [26] in the Bengali language, on the results of his epoch-making researches.

The Calcutta University next showed its belated recognition, by conferring on him the degree of D.Sc. *honoris causa*, in 1912.

And the Punjab University also showed its appreciation by inviting him, in 1913, to deliver a course of lectures on the results of his investigation.

PUBLIC [Pg_049] SERVICE COMMISSION

Dr. J. C. Bose was invited to give his evidence before the Royal Commission on the Public Services in India. With reference to the Method of Recruitment, he observed, in his written statement, as follows:—" ... I think that a high standard of scholarship should be the only qualification insisted on. Graduates of well-known Universities, distinguished for a particular line of study, should be given the preference. I think the prospects of the Indian Educational Service are sufficiently high to attract the very best material. In Colonial Universities they manage to get very distinguished men without any extravagantly high pay.... At present the recruitment in the Indian Educational Service is made in England and is practically confined to Englishmen. Such racial preference is, in my opinion, prejudicial to the interest of education. The best men available, English or Indian, should be selected impartially, and high scholarship should be the only test.... It is unfortunate that Indian graduates of European Universities who had distinguished themselves in a remarkable manner do not for one reason or other find facilities for entering the higher Educational Service.... I should like to add that these highly qualified Indians need only opportunities to render service which would greatly advance the cause of higher education.... If [Pg_050] promising Indian graduates are given the opportunity of visiting foreign Universities, I have no doubt that they would stand comparison with the best recruits that can be obtained from the West.... As teachers and workers it is an incontestable fact that Indian Officers have distinguished themselves very highly, and anything which discriminates between Europeans and Indians in the way of pay and prospects is most undesirable. A sense of injustice is ill-calculated to bring about that harmony which is so necessary among all the members of an educational institution, professors and students alike." [27] Pressing next for a high level of scholarship, in the Indian Educational Service, he wrote:—

"It has been said that the present standard of Indian Universities is not as high as that of British Universities, and that the work done by the former is more like that of the 6th form of the public schools in England. It is therefore urged that what is required for an Educational officer in the capacity to manage classes rather than high scholarship. I do not agree with these views. (1) There are Universities in Great Britain whose standards are not higher than ours; I do not think [Pg_051] that the Pass Degree even of Oxford or Cambridge is higher than the corresponding degree here (2) the standard of the Indian University is being steadily raised; (3) the standard will depend upon what the men entrusted with Educational work will make it. For these reasons it is necessary that the level of scholarship represented by the Indian Educational Service should be maintained very high." [28]

He then dwelt on what should be the aim of Higher Education in India and observed as follows:—

"... I think that all the machinery to improve the higher education in India would be altogether ineffectual unless India enters the world movement for the advancement of knowledge. And for this it is absolutely necessary to touch the imagination of the people so as to rouse them to give their best energies to the work of research and discovery, in which all the nations of the world are now engaged. To aim anything less will only end in lifeless and mechanical system from which the soul of reality has passed away." [28]

He was called, on the 18th December 1913, and was put to a searching examination by the Members [Pg_052] of the Royal Commission. The evidence that he gave is instinct with patriotism and is highly remarkable for its simplicity and directness about the things he said. To the Chairman (Lord Islington) he stated that he "favoured an arrangement by which Indians would enter the higher ranks of the service, either through the Provincial Service or by direct recruitment in India. The latter class of officers, after completing their education in India, should ordinarily go to Europe with a view to widening their experience. By this he did not wish to decry the training given in the Indian Universities, which produce some of the very best men, and he would not make the rule absolute. It was not necessary for men of exceptional ability to go to England in order to occupy a high chair. Unfortunately, on account of there being no openings for men of genius in the Educational Service, distinguished men were driven to the profession of Law. In the present condition of India a larger number of distinguished men were needed to give their lives to the education of the people.

"... The educational service ought to be regarded not as a profession, but as a calling. Some men were born to be teachers. It was not a question of race, of course; in order to have an efficient educational system, there must be an efficient organisation, but this should not be allowed to [Pg_053] become fossilised, and thus stand in the way of healthy growth.... A proportion of Europeans in the service, was needed, but only as experts and not as ordinary teachers. Only the very best men should be obtained from Europe and for exceptional cases. The general educational work should be done entirely by Indians, who understood the difficulties of the country much better than any outsider. He advocated the direct recruitment of Indians in India by the local Government in consultation with the

Secretary of State, rather than by the Secretary of State alone. Indians were under a great difficulty, in that they could not remain indefinitely in England after taking their degrees and being away from the place of recruitment their claims were overlooked. There was no reason why a European should be paid a higher rate of salary than an Indian on account of the distance he came. An Indian felt a sense of inferiority if a difference was made as regards pay. The very slight saving which Government made by differentiating between the two did not compensate for the feeling of wrong done. This feeling would remain even if the pay was the same, but an additional grant in the shape of a foreign service allowance was made to Europeans. All workers in the field of education should feel a sense of solidarity, because they [Pg_054] were all serving one greet cause, namely, education." [29]

Being asked by Sir Valentine Chirol, he said "If a foreign professor would not come and serve in India for the same remuneration as he obtained in his own country, he would certainly not force him to come." [29]

To Mr. Abdur Rahim he said: "Recruitment for the Educational Service should be made in the first place in India, if suitable men were available; but if not then he would allow the best outsiders to be brought in. In the present state of the country it would be very easy to fill up many of the chairs by selecting the best men in India. The aim of the universities should be to promote two classes of work—first, research; and, secondly, an all-round sound education...." [29]

In answer to questions of Mr. Madge, he said: "Any idea that the educational system of India was so far inferior to that of England, that Indians, who had made their mark, had done so, not because of the educational system of the country, but in spite of it, was quite unfounded. The standard of education prevailing in India was quite up to the mark of several British Universities. It was [Pg_055] as true of any other country in the world as of India that education was valued as a means for passing examination, and not only for itself, and there was no more cramming in India than elsewhere. The West certainly brought to the East a modern spirit, which was very valuable, but it would be dearly purchased by the loss of an honourable career for competent Indians in their own country. The educational system in India had in the past been too mechanical, but a turn for the better was now taking place and the Universities were recognising the importance of research work, and were willing to give their highest degrees to encourage it." [30]

To Mr. Fisher, he said that he "desired to secure for India Europeans who had European reputations in their different branches of study. If it was necessary to go outside India or England, to procure good men, he would prefer to go to Germany. This was the practice in America where they were annexing all the great intellects of Europe. He would like to see India entering the world movement in the advance and march of knowledge. It was of the highest importance that there should be an intellectual atmosphere in India. It would be of advantage if there were many Indians in the [Pg_056] Educational Service. For they came more in contact with the people, and influenced their intellectual activity. Besides, on retirement they would live in India, and their ripe experience would be at their countrymen's service." [31]

To Mr. Gokhale, he said that he "knew of three instances in which the Colonies had secured distinguished men on salaries which were lower than those given to officers of the Indian Educational Service. One was at Toronto, another was in New Zealand and the third at Yale University. The salaries on the two latter cases were #600 and #500 a year. The same held good as regards Japan. The facts there had been stated in a Government of India publication as follows: 'Subsequent to 1895 there were 67 professors recruited in Europe and America. Of these 20 came from Germany, 16 from England and 12 from the United States. The average pay was #384. In the highest Imperial University the average pay is #684. As soon as Japanese could be found to do the work, even tolerably well, the foreigner was dropped.' When he first started work in India, he found that there was no physical laboratory, or any grant made for a practical experimental course. He had [Pg_057] to construct instruments with the help of local mechanics, whom he had to train. All this took him ten years. He then undertook original investigation at his own expense. The Royal Society became specially interested in his work and desired to give him parliamentary grant for its continuation. It was after this that the Government of Bengal came forward and offered him facilities for research. In the Educational Service he would take men of achievement from any where; but men of promise he would take from his own country." [32]

To Sir Theodore Morison, he said: "There should be one scale of pay for all persons in the higher Educational Department. The rate of salary, Rs. 200 rising to Rs. 1,500 per month, was suitable subject to the proviso that a man of great distinction, instead of beginning at the lowest rate of pay, should start some where in the middle of the list, say, at Rs. 400 or Rs. 500. He would make no difference in regard to Europeans or Indians in that respect.... It would not be right for a great Government to grant a minimum of pay to Indian Professors and an extravagantly high pay to their [Pg_058] European Colleagues, for doing the same kind of work." [33]

To Mr. Gupta, he said that "He desired one Service, because he thought it was most degrading that certain man, although they were doing the same work should be classed in a Provincial Service, while others should be classed in an Imperial Service. The prospects of the members of the Provincial Service were not at all what they ought to be, and that was the reason why the best men were not attracted to it." [33]

FOURTH SCIENTIFIC DEPUTATION (1914-15)

Though the theories of Dr. Bose received acceptance from the leading scientific men of the Royal Society, yet Dr. Bose realised the necessity of bringing about a *general conviction* as to the truth of the identity of life-reactions in plant and in animal. So he looked for an opportunity of giving demonstration of his discoveries before the leading Scientific Societies of the World. And that opportunity came. The Royal Institution of Great Britain again invited him to deliver a 'Friday evening discourse' on the results of his new researches. The University of Oxford and Cambridge [Pg_059] also followed suit. The Government of India also showed their appreciation by sending him again on a Deputation for placing his discoveries before the Scientific world. He remained on deputation from the 3rd April 1914 to the 12th June 1915.

DR. BOSE IN EUROPE

Proceeding on his Deputation to England, Dr. Bose gave his first lecture, on the 20th May 1914, at Oxford,—where the late Sir John Burden Sanderson and his followers were the leaders of biological thought—in presence of very distinguished scientists. It was a grand success. Actual visualisation by physical demonstration of the results of his novel researches at once convinced those who were present. He next proposed to give a discourse on Plant Response before the University of Cambridge. The interest in this lecture became so very keen that the Botanical Department of Cambridge went to the length of importing soil from India to give the plants the most favourable conditions for exhibiting their specific reactions. At the lecture, the large Botanical Theatre became filled with scientific specialists, dons and advanced students, who followed with great attention the experiments with which he illustrated his discourse. He was greeted [Pg_060] with applause by the eminent scientists who thronged the lecture-theatre, at the end of every experiment. Sir Francis Darwin, the eminent botanist, in proposing a vote of thanks to Dr. Bose, said that 'he was filled with admiration, not only for the brilliancy of the work but for the convincing character of the experiments.' The scientists next assembled in great force, on the 29th May 1914, to hear the 'Friday Evening Discourse' of Dr. J. C. Bose on 'Plant Autographs and their Revelations,' at the Royal Institution, which was highly appreciated. At the end of the Discourse, Sir James Dewar, President of the Institution, gave an 'At Home' in honour of Dr. and Mrs. Bose. [34]

THE MAIDA VALE LABORATORY

The demonstrations of a far-reaching character which Dr. Bose gave evoked considerable public interest in England. His private laboratory at Maida Vale, in London, became the object of pilgrimage to the leading men of thought there. Sir William Crookes, the President of the Royal Society, came and became 'much impressed by the most ingenious and novel self-recording instruments.' Professor Starling, the author of the standard work on Physiology, and Professor [Pg_061] Oliver, the well-known Plant Physiologist, also became impressed by the delicacy and importance of Dr. Bose's work and methods. Professor Carveth Read, author of "Metaphysics of Nature," wondered how far the researches would profoundly affect the philosophical thoughts. Mr. Balfour, the ex-premier, became enthralled with what he saw. Professor James A. H. Murray, Editor of the 'Oxford New English Dictionary,' and Bernard Shaw, the famous dramatist, felt themselves attracted to the great Indian Scientist and came to pay their homage to him. Even Lord Crewe, the then Secretary of State for India, paid a visit to his laboratory and spoke warmly of the pride which he and the Government of India felt for his discoveries and of high gratification to him that India should once more make such contributions for the intellectual advancement of the world. The leading newspapers wrote eulogistically of his researches. The well-known scientific journal *Nature* devoted ten columns to an illustrated synopsis of his discoveries. Lord Hardinge, the then Viceroy, wrote a congratulatory letter to him—"It has been a source of immense gratification to the Viceroy to know that the foremost place in the special branch of research has been taken by one of India's most distinguished sons. The success you have won will only serve to stimulate your efforts and those of your pupils to other scientific investigations [Pg_062] which will redound still further to the honour of those who conduct them, and of India, the country of their birth." 35

From England Dr. Bose proceeded to the Continent, where his researches had already evoked keen interest.

On the 27th June 1914, he gave an address, illustrated with experiments, before the University of Vienna, which stands foremost in Biological researches. He was greeted with enthusiasm by the savants there. Some of the workers in plant physiology became so very much impressed with his demonstrations that they expressed a desire to be trained under him. Professor Molisch, the Director of the Pflanzen-physiologisches Institute of the Imperial University of Vienna, in proposing a vote of thanks, spoke highly of the great inspiration which the Viennese scientific men received from his discourse and dwelt on the indebtedness of Europe to India for the method of investigation initiated by Dr. Bose—method, which rendered it possible to prove deep into plant-life and bring forth results of which they could not hitherto dream. And the University of Vienna officially addressed the Secretary of State for India asking that special thanks of the University be conveyed to the Government of [Pg_063] India for the impetus given to them by Dr. Bose's visit. Dr. Bose was next to start for Germany on his scientific mission, and address the University of Strassburg, Leipzic, Halle, Berlin and Bonn and then attend the international congress at Munich, but, as the War broke out, he was compelled to come back to London. [36] On his way back, he gave a Discourse before the eminent scientific men in Paris.

On his return to London, medical men evinced great interest in his researches. Sir John Reid, President of the Royal Society of Medicine, and Sir Lauder Brunton, Physician of His Majesty the King Emperor, paid a visit to his laboratory to witness the action of drugs upon plants. Sir Lauder Brunton became of opinion that 'much light would be thrown on action of drugs on animals, by first observing their effects on plants.' As a result of this visit, Dr. Bose was invited to give an address to the Royal Society of Medicine in the beginning of winter. But, as the period of his Deputation was about to expire, the Society cabled to the Government of India for an extension, which was granted. Dr. Bose then delivered a lecture, before the Royal [Pg_064] Society of Medicine, on the 30th October 1914. The Royal Society of Medicine officially addressed the Secretary of State for India as follows: —

"... The lecture was one of the most successful we have had yet and evoked the keenest interest in the audience, Sir Lauder Brunton, Bt., and others taking part in the discussion, and warmly congratulating Prof. Bose and the Society on the value of his work. Since then I have received many expressions of appreciation that the Society was able to offer its fellows such an interesting demonstration of an entirely new departure in Biological Science." "At the invitation of the Psychological Society of London, Dr. Bose next delivered an interesting lecture on his theory of Memory Image." [37] He also gave an Address before the London Imperial college of Science.

DR. BOSE IN AMERICA

Dr. Bose's discoveries in the meantime evoked great interest in America. He was invited by several leading scientific bodies to come over there and acquaint them with the results of his wonderful researches. So he next went to America. "While in America, he was swamped with letters and telegrams for lecture engagements from Maine to [Pg_065] California" wrote Professor Sudhindra Bose M.A., Ph.D., of the Iowa University at that time, in the Modern Review. [38] "He has had so many calls for lectures from various Scientific societies, Colleges and Universities, that if he could speak twice a day and every day in the week, he could not hope to comply with all of those invitations in much less than a year." As he was in the United States, only for a few weeks, "he spoke before such learned bodies as the New York Academy of Sciences, the American Association for the Advancement of Science, the Brooklyn Institute of Arts and Science, the Philosophical Society of Philadelphia, and joint meeting of Academy of Science, the Botanical Society, and the Bureau of Plant Industry at Washington. Among the larger Universities, he gave addresses at Harvard, Columbia, Iowa, Illinois, Chicago, Michigan, Wisconsin.... Everywhere Dr. Bose has met with a very hearty welcome from the people of the American Republic. Even the Hon'ble Secretary of State, William Jennings Bryan, invited him to give a demonstration of his work at the State Department in Washington—an honour of unusual significance.... Dr. Bose has been made the subject of many magazine articles, newspaper editorials, cartoons and poems" [38] .... "The famous Smithsonian [Pg_066] Institute showed its high appreciation by submitting a report of Prof. Bose's work to the Congress. The Bureau of Plant Industry in Washington recognised his work on plant physiology as a very important contribution for the advancement of agriculture.... At the Harvard University his work has been received with high appreciation. President Stanley Hall, who is one of the leading psychologists of the day, has introduced Prof. Bose's work in the Post-graduate course of the Clarke University. His books have also been prescribed for physiological courses in different Universities in America, and in one of the leading Universities there, a special course of lectures is devoted to Prof. Bose's investigations on plant irritability...." [39]

The Columbia University, the largest in the United States, requested Dr. Bose to provide facilities in his Laboratory "for the reception of foreign students, who are desirous of familiarising themselves first hand with his apparatus and methods."

WHAT DR. BOSE SAW IN JAPAN

Dr. Bose then came back to India, in June 1915, *via* Japan. During his stay, in Japan, he acquainted himself with the efforts of the people and their aspirations towards a great future. He found that, [Pg_067] "in materialistic efficiency, which, in a mechanical era, is regarded as an index of civilisation, they have surpassed their German teachers. A few decades ago, they had no foreign shipping and no manufactures. But, within an incredibly short time, their magnificent lines of steamers have proved so formidable a competitor that the great American lines in the Pacific will soon be compelled to stop their sailings. Their industries again, through the wise help of the State and other adventitious aids, are capturing foreign markets. But far more admirable is their foresight to save their country from any embroilment with other nations with whom they want to live in peace. And they realise that any predominant interest of a foreign country in their trade or manufacture is sure to lead to misunderstanding and friction. Actuated by this idea, they have practically excluded all foreign manufactured articles by prohibitive tariffs." [40] "Is our country slow to realise the danger" asks Dr. Bose "that threatens her by the capture of her market and the total destruction of her industries? Does she not realise that it is helpless passivity that directly provokes aggression?... There is, therefore, no time to be lost and the utmost effort is demanded of the Government and the people for the revival of our industries...." [41]

A [Pg_068] PATRIOTIC CALL

"A very serious danger" continues Dr. Bose "is thus seen to be threatening the future of India, and to avert it will require the utmost effort of the people. They have not only to meet the economic crisis but also to protect the ideals of ancient Aryan civilisation from the destructive forces that are threatening it.... There is a danger of regarding the mechanical efficiency as the sole end of life; there is also the opposite danger of a life of dreaming, bereft of struggle and activity, the degenerating into parasitic habits of dependence. Only through the noble call of patriotism can our nation realise the highest ideals in thought and in action...." [42]

BACK TO INDIA

After his return to India, Dr. Bose attended the Indian Science Congress at Lucknow. He then attended the ceremony of the laying of the foundation stone of the Hindu University at Benares. On that occasion he delivered a masterly address. He said:—

"In tracing the characteristic phenomena of life from simple beginnings in that vast region which may be called unvoiced, as exemplified in the world of plants, to its highest expression in the animal [Pg_069] kingdom, one is repeatedly struck by the one dominant fact that in order to maintain an organism at the height of its efficiency something more than a mechanical perfection of its structure is necessary. Every living organism, in order to maintain its life and growth, must be in free communion with all the forces of the Universe about it.

"Further, it must not only constantly receive stimulus from without, but must also give out something from within, and the healthy life of the organism will depend on these two-fold activities of inflow and outflow. When there is any interference with these activities, then morbid symptoms appear, which ultimately must end in disaster and death. This is equally true of the intellectual life of a Nation. When through narrow conceit a Nation regards itself self-sufficient and cuts itself from the stimulus of the outside world, then intellectual decay must inevitably follow.

"So far as regards the receptive function. Then there is another function in the intellectual life of a Nation, that of spontaneous flow, that going out of its life by which the world is enriched. When the Nation has lost this power, when it merely receives, but cannot give out, then its healthy life is over, and it sinks into a degenerate existence, which is purely parasitic.

"How can our Nation give out of the fulness of [Pg_070] the life that is in it, and how can a new Indian University help in the realisation of this object? It is clear that its power of directing and inspiring will depend on its world status. This can be secured to it by no artificial means, nor by any strength in the past....

"This world status can only be won by the intrinsic value of the great contributions to be made by its own Indian scholars for the advancement of the world's knowledge. To be organic and vital our new University must stand primarily for self-expression and for winning for India a place she has lost. Knowledge is never the exclusive possession of any particular race, nor does it recognise geographical limitations. The whole world is interdependent, and a constant stream of thought has been carried out throughout the ages enriching the common heritage of mankind. Although science was neither of the East nor of the West but international, certain aspects of it gained richness by reason of their place of origin." [43]

OUTCOME OF THE SCIENTIFIC MISSION

The scientific mission of Dr. Bose to the West was a great success. The very convincing character of the demonstrations that he gave, before the [Pg_071] leading Scientific Societies of the world, with his newly invented Resonant Recorder and other delicate instruments, secured a world-wide acceptance of his theories and results. Not only that. He secured also a recognition from the leading thinkers of "that trend of thought which led him unconsciously to the dividing frontiers of different sciences and shaped the course of his work." [44] It has come to be recognised that "India through her habit of mind is peculiarly fitted to realise the idea of unity and to see in the phenomenal world an orderly universe," to realise that "there can be but one truth, one Science which includes all other branches of knowledge," [44] and that the store of world's knowledge would be incomplete without India's special contribution to it. Thus he has raised India in the estimation of the intellectual world.

RETIREMENT FROM GOVERNMENT SERVICE

Dr. Bose reached the age limit of 55 on the 29th November 1913 but he was granted an extension till the 13th September 1915. The period of his extension having expired, he retired from the Professorship in the Presidency College after 31 years of service. The Governing Body of the College, [Pg_072] however, "in recognition of his eminent services to Science and Presidency College," appointed him *honoris causa* Emeritus Professor of the College. His duties as a member of the staff ceased. But he was given facilities to continue his work in the Physical Laboratory of the College. [45]

FURTHER RECOGNITION

After his retirement, the Secretary of State, who had already been impressed with the high value of his researches, sanctioned a recurring grant of Rs. 30,000 a year (for him and his assistants) for 5 years and a non-recurring grant of Rs. 25,000 (for equipment) for continuation of his original work.... And, in further recognition of his valuable scientific work, the Government conferred on him a Knighthood, on the 1st January 1917. It may, however, be mentioned that this high honour has been bestowed for the first time on an Indian for his original work in Science.

FEELS THE NECESSITY FOR THE FOUNDATION OF AN INSTITUTE

Relieved of the trammels of service, Dr. Bose felt the necessity for realising a dream that wove a network round his wakeful life for years past—for [Pg_073] establishing an Institute—a Study and Garden of Life—where the creepers, plants and trees would be played upon by their natural environment and would transcribe in their own script the history of their experience, where "the student would watch the panorama of life" and, "isolated from all distractions, would learn to attune himself with Nature and to see how community throughout the great ocean of life outweighs apparent the dissimilarity," and where "the genius of India would find its true blossoming," where the "synthetical intellectual methods of the East would co-operate with the analytical methods of the West," and whence would emanate a rich and peculiar current of thought and to which would be attracted votaries from all lands. [46]

THE BOSE INSTITUTE

Though the realisation of such a glorious Institute would not be effected through one life or one fortune, he wanted to accomplish something—something, so far as it lay in his power. So he proceeded to build and equip an Institute—the "Bose Institute"—at a cost of about 5 lakhs, the entire savings of his lifetime. While it was being constructed Their Excellencies the Viceroy and the [Pg_074] Governor of Bengal paid a visit to Dr. Bose's private laboratory. On the 30th November 1917—the anniversary of his sixtieth birthday—he dedicated the Institute to the Nation, for the progress of Science and for the Glory of India.

THE AIMS OF THE INSTITUTE

In this Institute, Dr. Bose intends to go on with "the further and fuller investigation of the many and ever-opening problems of the nascent science which includes both Life and None Life" and wants to train up a devoted band of workers, with the Sanyasin mind, who would keep alive the flame kindled by him, and who, by acute observation and patient experiment would "wring out from Nature some of her most jealously guarded secrets" and who would thus lead to the establishment of a great Indian School of Science and to the "building of the greater India yet to be." There would be no academic limitation here to the widest possible diffusion of knowledge. The facilities of the Institute would be available to workers from all countries and there would be no desecration of knowledge here by its utilisation for personal gain—no patent would be taken of the discoveries here made. The high aim of a great Seat of Learning would be sought to be maintained here. The lectures here given would not be mere repetitions, second-hand knowledge but would announce [Pg_075] for the first time to the world the new discoveries here made. [47]

The efforts of Dr. Bose have also animated our countrymen. Maharaja Sir Manindra Chandra Nandy of Kasimbazar has made a gift of two lakhs to the Institute. Mr. S. R. Bomanji has given one lakh. Mr. Moolraj Khatao has endowed the Institute with two lakh and a quarter. Other contributions are still pouring in.

A GREAT 'SADHAK'

With a true *Sanyasin* spirit, Dr. Bose applied himself to the study of Nature. His ardour was ever compassable. Even the limitations of the senses would hardly fetter him in his explorations in the regions of the Unknown. He expended the range of perception by means of wonderfully sensitive instrumental devices. By acute observations and patient experiment he wrung out from Nature some of her most jealously guarded secrets in the realm of Electric Radiation, which "literally filled with wonder and admiration" the greatest scientist of the age. Allurements of great material prospects—which might lead him to the path of immense fortune—came to him, in the shape of the patents of his inventions. But they had no attraction for [Pg_076] him. In utter disregard of all worldly advancement, he continued in his pursuit of knowledge.

In pursuit of his investigations on Electric Radiation, he was unconsciously led into the border region of Physics and Physiology. He caught a glimpse of ineffable wonder that remained hidden behind the view. He attempted to lift the veil. And, at once, difficulties presented themselves one after another. An unfamiliar caste in the domain of Science got offended. He was asked not to encroach on the special preserve of the Physiologists and, as he did not pay any heed to the warning, misrepresentations began. Even the evidence of his supersensitive appliances failed to convince many. And the Royal Society withheld publication of his researches. He was recompensed with ridicule and reviling. The limited facilities that he had in the prosecution of his researches were in danger of being withdrawn. But he had a burning Faith in the Vision and was not to be boggled at with these difficulties. He became stronger in his determination. Realising an inner call, he dedicated himself for the establishment of the truth underlying his Faith. He cast his life, as an offering, regarding success and failure as one, and engaged himself in a protracted struggle to get behind the deceptive seeming into the reality that remained unseen. After years of sustained efforts, [Pg_077] he succeeded in overcoming almost insuperable difficulties in the way of the realisation of the great dream of his life. The closed doors at last opened, and the seemingly impossible became possible. The secret of the plant world stood revealed by the autographs of the plants themselves. "It was when I came upon the mute witness of these self-made records," said Sir J. C. Bose, when he stood before the Royal Institution "and perceived in them one phase of a pervading unity that bears within it all things: the mote that quivers in ripples of light, the teeming life upon our earth, and the radiant suns that shine above us — it was then that I understood for the first time a little of that message proclaimed by my ancestors on the banks of the Ganges thirty centuries ago."

"They who see but one in all the changing manifestations of this universe, unto them belongs Eternal Truth — unto none else, unto none else." [48]

The Rishis of ancient India, by their intense Yoga, realised the One in the Many. But Sir Jagadis Chandra, by rigorous experimental demonstration, realised a Unity amidst Diversity. He perceived that "there was no such thing as brute [Pg_078] matter, but that spirit suffused matter in which it was enshrined." [49]

EFFECT OF HIS WORK

It is impossible to estimate the effect of his epoch-making researches. The psychic stone flung by him into the pool of physical botany, has made the ripples run in so many directions. There have been produced "unexpected revelations in plant life, foreshadowing the wonders of the highest animal life." And there "have opened out very extended regions of inquiry in Physics, in Physiology, in Medicine, in Agriculture and even in Psychology. Problems, hitherto regarded as insoluble, have now been brought within the sphere of experimental investigation."

Sir J.C. Bose has not only extended the distant boundaries of Science, but, by his peculiarly Indian contribution, has secured a recognised place for India and has revived a hope in the Indian mind that India may yet regain a place among the intellectual nations of the world. Men like him are rare not only in India but rare any where in the world. May he live long!

Footnotes

[1] Vide 'History of a Failure that was great' — Modern Review, Vol. XXI, p. 221.

[2] Vide 'History of a Failure that was great' — Modern Review. Vol. XXI p. 221.

[3] *Vide* 'History of a failure that was great' — Modern Review, Vol. XXI, p 221.

[4] 'History of a Failure that was great' — Modern Review. Vol, XXI, p. 221.

[5] Convocation Address, dated 2nd March 1907, delivered by Sir Ashutosh Mookerjea.

[6] Vide Evidence of Dr. J. C. Bose before the Public Services Commission, — Vol. XX, p. 136.

[7] Address to the Royal Society by its President, Sir Benjamin Brodie, 30th November 1859.

[8] 1 metre = 39.4 inches.

[9] Encyclopfdia Britannica, 11th Edition, Vol IX, p. 206.

[10] Encyclopfdia Britannica, 11th Edition, Vol. IX, p. 206.

[11] See 'History of a Discovery' — Modern Review, Vol. XVIII, p. 693.

[12] See 'Voice of Life' — Modern Review, Vol. XII, p. 590.

[13] Vide 'History of a Discovery' — Modern Review, Vol. XVIII, p. 694.

[14] Response in Living and Non-Living, p. 191.

[15] See 'Voice of Life' — Modern Review, Vol. XXII, p. 588.

[16] See 'History of a Discovery'—Modern Review, Vol. XVIII, p. 694.

[17] Vide 'Voice of Life'—Modern Review, Vol. XXII, p. 592.

[18] See 'Voice of Life'—Modern Review, Vol. XXII, p. 592.

[19] Vide 'Voice of Life'—Modern Review, Vol. XXII, p. 592.

[20] Vide 'History of a Discovery'—Modern Review, Vol. XVIII, p. 694.

[21] Cf. Preface to 'Comparative Electro-Physiology' p. IX.

[22] Vide 'Comparative Electro-Physiology' pp. 732-733.

[23] Vide 'Memory Image and its Revival,' Sir J. C. Bose—Modern Review, Vol. XXIV, p. 447.

[24] Sri Sermon on "Prayer" delivered by Keshub Chunder Sen at the Prarthana Samaj, Bombay, on March 26, 1868.

[25] See 'Voice of Life'—Modern Review, Vol. XXII, p. 588.

[26] Vide Modern Review Vol. XI, p. 539.

[27] Vide Appendix to the Report of the Royal Commission on the Public Services in India, Vol. XX, p. 135-136.

[28] Vide Appendix to the Report of the Royal Commission on the Public Services in India, Vol. XX, p. 135.

[29] Vide Appendix to the Report of the Royal Commission on the Public Services in India, Vol. XX, p. 136

[30] Vide Appendix to the Report of the Royal Commission on the Public Services in India, Vol. XX, p. 137.

[31] Vide Appendix to the Report of the Royal Commission on the Public Services in India, Vol. XX, p. 137.

[32] Vide Appendix to the Report of the Royal Commission on the Public Services in India, Vol. XX, p. 137.

[33] Vide Appendix to the Report of the Royal Commission on the Public Services in India, Vol. XX, p. 139.

[34] Vide Modern Review—Vol. XVI, pp. 16, 118, 120.

[35] Vide Modern Review, Vol. XVI, pp. 120, 121, 126.

[36] Vide Modern Review, Vol. XVII, P. 559.

[37] Vide Modern Review, Vol. XVI, p. 246.

[38] Vide Modern Review, Vol. XVII, p. 559.

[39] Vide Modern Review, Vol. XVIII, p. 1.

[40] Vide Modern Review, Vol. XVIII. p. 214.

[41] Vide Modern Review, Vol. XVIII. p. 215.

[42] Vide Modern Review, Vol. XVIII, p. 215.

[43] Vide Modern Review, Vol. XIX, p. 277.

[44] Vide 'Voice of Life' — Modern Review, Vol. XXII, p. 591.

[45] Presidency College Magazine, Vol. II, p. 335.

[46] Presidency College Magazine, Vol. II, p, 335.

[47] Vide 'Voice of Life' — Modern Review, XXII, p. 590.

[48] Vide 'Voice of Life' — Modern Review Vol XXII, p. 590.

[49] Vide Modern Review, Vol. XXI, p. 343.

LITERATURE AND SCIENCE

The following is a substance of the Address delivered in Bengali by Prof. J. C. Bose, on the 14th April 1911, as the President of the Bengal Literary Conference, which met in the Easter of 1911 at Mymensing.

In this Literary Congress it would appear that you have interpreted Letters in no exclusive sense. We are not met to discuss the place that literature is to hold in the gospel of beauty. Rather are we set upon conceiving of her in larger ways. To us to-day literature is no mere ornament, no mere amusement. Instead of this, we desire to bring beneath her shadow all the highest efforts of our minds. In this great communion of learning, this is not the first time that a scientific man has officiated as priest. The chair which I now occupy has already been held by one whom I love and honour as friend and colleague, and glory in our countryman, Praphulla Chandra Ray. In honouring him, your Society has not only done homage to merit, but has also placed before our people a lofty and inclusive ideal of literature.

[Pg_080] You are aware that in this West, the prevailing tendency at the moment is, after a period of synthesis, to return upon the excessive sub-division of learning. The result of this specialisation is rather to accentuate the distinctiveness of the various sciences, so that for a while the great unity of all tends perhaps to be obscured. Such a caste system in scholarship, undoubtedly helps at first, in the gathering and classification of new material. But if followed too exclusively, it ends by limiting the comprehensiveness of truth. The search is endless. Realisation evades us.

The Eastern aim has been rather the opposite, namely, that in the multiplicity of phenomena, we should never miss their underlying unity. After generations of this quest, the idea of unity comes to us almost spontaneously, and we apprehend no insuperable obstacle in grasping it.

I feel that here in this Literary Congress, this characteristic idea of unity has worked unconsciously. We have never thought of narrowing the bounds of literature by a jealous definition of its limits. On the contrary, we have allowed its empire to extend. And you have felt that this could be adequately done only, if in one place you could gather together all that we are seeking, all that we are thinking, all that we are examining. And for this you have to-day invited those who sing along [Pg_081] with those who meditate, and those who experiment. And this is why, though my own life has been given to the pursuit of science, I had yet no hesitation in accepting the honour of your invitation.

POETRY AND SCIENCE

The poet, seeing by the heart, realises the inexpressible and strived to give it expression. His imagination soars, where the sight of others fails, and his news of realm unknown finds voice in rhyme and metre. The path of the scientific man may be different, yet there is some likeness between the two pursuits. Where visible light ends, he still follows the invisible. Where the note of the audible reaches the unheard, even there he gathers the tremulous message. That mystery which lies behind the expressed, is the object of his questioning also; and he, in his scientific way, attempts to render its abstruse discoveries into human speech.

This vast abode of nature is built in many wings, each with its own portal. The physicist, the chemist, and the biologist entering by different doors, each one his own department of knowledge, comes to think that this is his special domain, unconnected with that of any other. Hence has arisen our present rigid division of phenomena, into the worlds of the inorganic, vegetal, and sentient. But [Pg_082] this attitude of mind is philosophical, may be denied. We must remember that all enquiries have as their goal the attainment of knowledge in its entirety. The partition walls between the cells in the great laboratory are only erected for a time to aid this search. Only at that point where all lines of investigation meet, can the whole truth be found.

Both poet and scientific worker have set out for the same goal, to find a unity in the bewildering diversity. The difference is that the poet thinks little of the path, whereas the scientific man must not neglect. The imagination of the poet has to be unrestricted. The intuitions of emotion cannot be established by rigid proof. He has, therefore, to use the language of imagery, adding constantly the words 'as if.'

The road that the scientific man has to tread is on the other hand very rugged, and in his pursuit of demonstration he must pay a severe restraint on his imagination. His constant anxiety is lest he should be self-deceived. He has, therefore, at every step to compare his own thought with the external fact. He has remorselessly to abandon all in which these are not agreed. His reward is that he gets, however little is certain, forming a strong foundation for what is yet to come. Even by this path of self-restraint and verification, however, he is [Pg_083] making for a region surpassing wonder. In the range of that invisible light, gross objects cease to be a barrier, and force and matter become less aesthetic. When the veil is suddenly lifted, upon the vision hitherto unsuspected, he may for a moment lose his accustomed self-restraint and, exclaim "not 'as if' — but the thing itself!"

INVISIBLE LIGHT.

In illustration of this sense of wonder which links together poetry and science, let me allude briefly to a few matters that belong to my own small corner in the great universe of knowledge, that of light invisible and of life unvoiced. Can anything appeal more to the imagination than the fact that we can detect the peculiarities in the internal molecular structure of an opaque body by means of light that is itself invisible? Could anything have been more unexpected than to find that a sphere of China-clay focuses invisible light more perfectly than a sphere of glass focuses the visible; that in fact, the refractive power of this clay to electric radiation is at least as great as that of the most costly diamond to light? From amongst the innumerable octaves of light, there is only one octave, with power to excite the human eye. In reality, we stand, in the midst of a luminous ocean, almost blind! The little that we can see is nothing, compared to the vastness of [Pg_084] that which we cannot. But it may be said that out of the very imperfection of his senses man has been able, in science, to build for himself a raft of thought by which to make daring adventure on the great seas of the unknown.

UNVOICED LIFE.

Again, just as, in following up light from visible to invisible, our range of investigation transcends our physical sight, so also does our power of sympathy become extended, when we pass from the voiced to the unvoiced, in the study of life: Is there then any possible relation between our own life and that of the plant world? That there may be such a relation, some of the foremost of scientific men have denied. So distinguished a leader as the late Burdon-Sanderson declared that the majority of plants were not capable of giving any answer, by either mechanical or electrical excitement, to an outside stock. Pfeffer, again, and his distinguished followers, have insisted that the plants have neither a nervous system, nor anything analogous to the nervous impulse of the animal. According to such a view, that two streams of life, in plant and animal, flow side by side, but under the guidance of different laws. The problems of vegetable life are, it must be said, extremely obscure, and for the penetrating [Pg_085] of that darkness we have long had to wait for instruments of a superlative sensitiveness. This has been the principal reason for our long clinging to mere theory, instead of looking for the demonstration of facts. But to learn the truth we have to put aside theories, and rely only on direct experiment. We have to abandon all our preconceptions, and put our questions direct, insisting that the only evidence we can accept is that which bears the plant's own signature.

How are we to know what unseen changes take place within the plant? If it be excited or depressed by some special circumstance, how are we, on the outside, to be made aware of this? The only conceivable way would be, if that were possible, to detect and measure the actual response of the organism to a definite external blow. When an animal receives an external shock it may answer in various ways if it has voice, by a cry; if it be dumb, by the movement of its limbs. The external shock is a stimulus; the answer of the organism is the response. If we can find out the relation between this stimulus and the response, we shall be able to determine the vitality of the plant at that moment. In an excitable condition, the feeblest stimulus will evoke an extraordinarily large response: in a depressed state, even a strong stimulus evokes only a feeble response; and lastly, when [Pg_086] death has overcome life, there is an abrupt end of the power to answer at all.

We might therefore have detected the internal condition of the plant, if, by some inducement, we could have made it write down its own responses. If we could once succeed in this apparently impossible task we should still have to learn the new language and the new script. In a world of so many different scripts, it is certainly undesirable to introduce a new one! I fear the Uniform Script Association will cherish a grievance against us for this. It is fortunate however that the plant-script bears, after all, a certain resemblance to the Devanagari—inasmuch as it is totally unintelligible to any but the very learned!

But there are two serious difficulties in our path; first, to make the plant itself consent to give its evidence; second, through plant and instrument combined, to induce it to give it in writing. It is comparatively easy to make a rebellious child obey: to extort answers from plants is indeed a problem! By many years of close contiguity, however, I have come to have some understanding of their ways. I take this opportunity to make public confession of various acts of cruelty which I have from time to time perpetrated on unoffending plants, in order to compel them to give me answers. For this purpose, I have devised various forms of torment,— [Pg_087] pinches simple and revolving, pricks with needles, and burns with acids. But let this pass. I now understand that replies so forced are unnatural, and of no value. Evidence so obtained is not to be trusted. Vivisection, for instance, cannot furnish unimpugnable results, for excessive shock tends of itself to make the response of a tissue abnormal. The experimental organism must therefore be subjected only to moderate stimulation. Again, one has to choose for one's experiment a favourable moment. Amongst plants, as with ourselves, there is, very early in the morning, especially after a cold night, certain sluggishness. The answers, then, are a little indistinct. In the excessive heat of midday, again, though the first few answers are very distinct, yet fatigue soon sets in. On a stormy day, the plant remains obstinately silent. Barring all these sources of aberration, however, if we choose our time wisely, we may succeed in obtaining clear answers, which persist without interruption.

It is our object, then, to gather the whole history of the plant, during every moment between its birth and its death. Through how many cycle of experience it has to pass! The effects on it of recurring light and darkness; the pull of the earth, and the blow of the storm; how complex is the concatenation of circumstances, how various are the shocks, and how multiplex are the replies which we have [Pg_088] to analyse! In this vegetal life which appears so placid and so stationary, how manifold are the subtle internal reactions! Then how are we to make this invisible visible?

THE DIARY OF THE PLANT.

The little seedling we know to be growing, but the rate of its growth is far below anything we can directly perceive. How are we to magnify this so as to make it instantly measurable? What are the variations in this infinitesimal growth under external shock? what changes are induced by the action of drugs or poisons? will the action of poison change with the dose? Is it possible to counteract the effect of one by another?

Supposing that the plant does not give answers to external shock, what time elapses between the shock and the reply? Does this latent period undergo any variation with external conditions? Is it possible to make the plant itself write down this excessively minute time-interval?

Next, does the effect of the blow given outside reach the interior of the plant? If so, is there anything analogous to the nerve of the animal? If so, again, at what rate does the nervous impulse travel the plant? By what favourable circumstances will this rate of transmission become enhanced, and [Pg_089] by what will be retarded or arrested? Is it possible to make the plant itself record this rate and its variations? Is there any resemblance between the nervous impulse in plants and animals? In the animal there are certain automatically pulsating tissues like the heart. Are there any such spontaneously beating tissues in a plant? What is the meaning of spontaneity? And lastly, when by the blow of death, life itself is finally extinguished, will it be possible to detect the critical moment? And does the plant then exert itself to make one overwhelming reply, after which response ceases altogether? Its autobiography can only be regarded as complete, if, with the help of efficient instruments, all these questions can be answered by it, so as to form the different chapters.

"If the plant could have been made thus to keep its own diary, then the whole of its history might have been recovered!" But words like these are born of day dreams merely. Vague imaginings of this kind may furnish much gratification to an idle life. When, awaking from these pleasant dreams of science, we seek to actualise the conditions imposed by them, we find ourselves face to face with a dead wall. For the doorway of nature's court is barred with iron, and through it can penetrate no mere cry of childish petulance. It is only by the gathered force of many years of concentration, that the gate [Pg_090] can be opened, and the seeker enter to explore the secrets that have baffled him so long.

DIFFICULTIES OF RESEARCH IN INDIA.

We often hear that without a properly equipped laboratory, higher research in this country is an absolute impossibility. But while there is a good deal in this, it is not by any means the whole truth. If it were all, then from these countries where millions have been spent on costly laboratories, we should have had daily accounts of new discoveries. Such news we do not hear. It is true that here we suffer from many difficulties, but how does it help us, to envy the good fortune of others? Rise from your depression! Cast off your weakness! Let us think, "In whatever condition we are placed, that is the true starting-point for us." India is our working-place, and all our duties are to be accomplished here, and nowhere else. Only he who has lost his manhood need repine.

In carrying out research, there are other difficulties, besides the want of well-equipped laboratories. We often forget that the real laboratory is one's own mind. The room and the instruments only externalise that. Every experiment has first to be carried out in that inner region. To keep the mental vision clear, great struggles have to be [Pg_091] undergone. For its clearness is lost, only too easily. The greatest wealth of external appliances is of no avail, where there is not a concentrated pursuit, utterly detached from personal gain. Those whose minds rush hither and thither, those who hunger for public applause instead of truth itself, by them the quest is not won. To those on the other hand, who do long for knowledge itself, the want of favourable conditions does not seem the principle obstacle.

In the first place, we have to realise that knowledge for the sake of knowledge is our aim, and that the world's common standard of utility have no place in it. The enquirer must follow where he is led, holding the quiet faith that things which appear to-day to be of no use, may be of the highest interest to-morrow. No height can be climbed, without the hewing of many an unremembered step! It is necessary, then, that the enquirer and his disciples should work on ceaselessly, undeterred by years of failure, and undistracted by the thunder of public applause. We may one day come to realise that India in the past has shared her knowledge with the world, and we may ask ourselves, is that destiny now ended for us? Are we of to-day to be debtors only? Perhaps when we have once felt this, a new Nalanda may arise.

THE PHYTOGRAPH

I was speaking of the need of various delicate instruments—phytographs, as I shall call them—for the automatic record of the plant's responses. What was, ten years ago, a mere aspiration, has now after so many years of effort, become actual fact. It is unnecessary to tell here of many a fruitless and despairing attempt. Nor shall I trouble you with any account of intricate mechanism. I need only say that with the aid of different types of apparatus, it is now possible for all the responsive activities of the plant to be written down. For instance, we can make an instantaneous record of the growth and its variations, moment by moment. Scripts can be obtained of its spontaneous movement. And a recording arm will demorcate the line of life from that of death. The extreme delicacy of one of these instruments will be understood, when it is said that it measures and records a time-interval so short as one-thousandth part of a second!

It has been supposed that instruments for research of this delicacy and precision, were only possible of construction in the best scientific manufactories of Europe. It will therefore be regarded as interesting and encouraging to know that every one of these has been executed entirely in India, by Indian workmen and mechanicians.

With perfect instruments at our disposal, we may [Pg_093] proceed to describe a few amongst the many phenomena which now stand revealed. But before this, it is necessary to deal briefly with the superstition that has led to the division of plants into sensitive and insensitive. By the electrical mode of investigation, it can be shown that not only Mimosa and the like, but all plants of all kinds are sensitive, and give definite replies to impinging stimuli. Ordinary plants, it is true, are unable to give any conspicuous mechanical indication of excitement. But this is not because of any insensitiveness, but because of equal and antagonistic reactions which neutralise each other. It is possible, however, by employing appropriate means, to show that even ordinary plants give mechanical replies to stimulus.

THE DETERMINATION OF THE LATENT PERIOD

When an animal is struck by a blow, it does not respond at once. A certain short interval elapses between the incidence of the blow, and the beginning of the reply. This lost time is known as the latent period. In the leg of a frog, the latent period according to Helmwoltz, is about one-hundredth of a second. This latent period, however, undergoes appropriate variation with changing external conditions. With feeble stimulus, it has a definite value, which, with an excessive blow, is much [Pg_094] shortened. In the cold season, it is relatively long. Again, when we are tired our perception time, as we may call it, may be greatly prolonged. Every one of these observations is equally applicable to the perception time of the plant. In Mimosa, in a vigorous condition, the latent period is six one hundredth of a second, that is to say, only six times its value in an energetic frog! Another curious thing is that a stoutish tree will give its response in a slow and lordly fashion, whereas a thin one attains the acme of its excitement in an incredibly short time! Perhaps some of us can tell from our own experience whether similar differences obtain amongst human kind or not? The plant's latent period in our cold weather may be almost doubled. Ordinarily speaking it takes *Mimosa* about fifteen minutes to recover from a blow. If a second blow be given, before the full recovery of its equanimity, then the plant becomes fatigued, and its latent period is lengthened. When over-fatigued, it may temporarily lose its power of perception altogether, what this condition is like, my audience is only too likely to realise, at the end of my long address!

THE RELATION BETWEEN STIMULUS AND RESPONSE

According to varying circumstances, the same blow will evoke responses of different amplitudes. [Pg_095] Early in the morning, after the prolonged inactivity of a cold night, we find the plant inclined to be lethargic, and its first answers correspondingly small. But as blow after blow is delivered, this lethargy passes off, and the replies become stronger and stronger. A good way to remove this lethargy quickly, is to give the plant a warm bath. In the heat of the midday, this state of things is reversed. That is to say, after giving vigorous replies the plant becomes fatigued, and its responses grow smaller. This fatigue passes off, however, on allowing it a period of rest. On increasing the intensity of the impinging stimulus, the response also increases. But a limit is attained, beyond which response can no longer be enhanced. Again, just as the pain of a blow persists longer with ourselves, in winter than in summer, so the same holds good of the reaction of the plant also. For instance, in summer it takes *Mimosa* ten to fifteen minutes to recover from a blow, whereas in winter the same thing would take over half an hour. In all this, you will recognise the similarity between human response and that of the plant.

SPONTANEOUS PULSATION

In certain tissues, a very curious phenomenon is observed. In man and other animals, there are tissues which beat, as we say, spontaneously. As [Pg_096] long as life lasts, so long does the heart continue to pulsate. There is no effect without a cause. How then was it that these pulstations [became spontaneous? To this query, no fully satisfactory answer has been forthcoming. We find, however, that similar spontaneous movements are also observable in plant tissues, and by their investigation the secret of automatism in the animal may perhaps be unravelled.

Physiologists, in order to know the heart of man, play with those of the frog and tortoise. "To know the heart," be it understood, is here meant in a purely physical, and not in a poetic sense. For this it is not always convenient to employ the whole of the frog. The heart is therefore cut out, and make the subject of experiments, as to what conditions accelerate, and what retard, the rate and amplitude of its beat. When thus isolated, the heart tends of itself to come to a standstill, but if, by means of fine tubing, it be then subjected to interval blood pressure, its beating will be resumed, and will continue uninterrupted for a long time. By the influence of warmth, the frequency of the pulsation may be increased, but its amplitude diminished. Exactly the reverse is the effect of cold. The natural rhythm and the amplitude of the pulse undergo appropriate changes, again, under the action of different drugs. Under either, the [Pg_097] heart may come to a standstill, but on blowing this off the beat is renewed. The action of chloroform is more dangerous, any excess in the dose inducing permanent arrest. Besides these, there are poisons also which arrest the heart beat, and a very noticeable fact in this connection is, that some stop in a contracted, and others in a relaxed condition. Knowing these opposed effects, it is sometimes possible to counteract the effect of one poison by administering another.

I have thus briefly stated some of the most important phenomena in connection with spontaneous movements in animal tissues. Is it possible that in plants also any parallel phenomena might be observed? In answer to this question, I may say that I have found numerous instances of automatic movements in plants.

RHYTHMIC PULSATIONS IN DESMODIUM

The existence of such spontaneous movements can easily be demonstrated, by means of our Indian *Bon charal*, the telegraph plant, or Desmodium gyrans, whose small leaflets dance continually. The popular belief that they dance in response to the clapping of the hands is quite untrue. From readings of the scripts made by this plant, I am in a position to state that the automatic movements of both plants and animals are guided by laws which are identical.

[Pg_098] Firstly, when, for convenience of experiment, we cut off the leaflet, its spontaneous movements, like those of the heart, come to a stop. But if we now subject the isolated leaflet, by means of a fine tube, to an added internal pressure of the plant's sap, its pulsations are renewed, and continue uninterrupted for a very long time. It is found again that the pulsation frequency is increased under the action warmth, and lessened under cold, increased frequency being attended by diminution of amplitude and *vice versa*. Under either, there is temporary arrest, revival being possible when the vapour is blown off. More fatal is the effect of chloroform. The most extraordinary parallelism, however, lies in the fact that those poisons which arrest the beat of the heart in a particular way, arrest the plant—pulsation also in a corresponding manner. I have thus been able to revive a leaflet poisoned by the application of one, with a dose of a counteracting poison.

Let us now enquire into the causes of these automatic movements so-called. In experimenting with certain types of plant-tissues, I find that an external stimulus may not always evoke an immediate reply. What happens, then, to the incident energy? It is not really lost, for these particular plant-tissues have the power of shortage. In this way, energy derived in various ways from without—as light, [Pg_099] warmth, food, and so on—is constantly being accumulated, when a certain point is reached, there is an overflow, and we call this overflow spontaneous movement. Thus what we call automatic is really an overflow of what has previously been stored up. When this accumulated energy is exhausted, then there is also an end of spontaneous movements. By abstracting its stored-up heat—through the application of cold water—we can bring to a stop the automatic pulsations of Desmodium. But on allowing a first accession of heat from outside, these pulsations are gradually restored.

In the matter of these so-called spontaneous activities of the plant, I find that there are two distinct types. In one, the overflow is initiated with very little storage, but here the unusual display of activity soon comes to a stop. To maintain such specimens in the rhythmic condition, constant stimulation from outside is necessary. Plants of this type are extremely dependent on outside influences, and when such sources of stimulus are removed, they speedily come to an inglorious stop. *Kamranga* or *Averrhoa* is an example of this kind. In the second type of automatic plant activity I find that long continued storage is required, before an overflow can begin. But in this case, the spontaneous outburst is persistent and of long duration, even when the plant is deprived of any immediately [Pg_100] exciting cause. These, therefore, are not so obviously dependent as the others on the sunshine of the world. Our telegraph-plant, *Desmodium* or *Bon charal*, is an example of this.

It appears to me that we have here a suggestive parallel to certain phenomena with which this audience will surely prove more familiar than I, namely, the facts of literary inspiration. For the attainment of this exalted condition, also, is it not necessary to have previous storage, with a consequent bubbling overflow? Certain indications incline me to suspect that perhaps in this also we have an example of so-called spontaneity, or automatic responsiveness. If this be so, aspirants, to the condition might well be asked to decide in whose footsteps they will choose to tread — those of *Kamranga*, with its dependence on outside influences, and inevitably ephemeral activity, or those of *Bon charal*, with its characteristic of patient long enduring accumulation of forces, to find uninterrupted and sustained expression.

THE PLANT'S RESPONSE TO THE SHOCK OF DEATH

A time comes when, after one answer to a supreme shock, there is a sudden end of the plant's power to give any response. This supreme shock is the shock of death. Even in this crisis, there is no immediate [Pg_101] change in the placid appearance of the plant. Drooping and withering are events that occur long after death itself. How does the plant then, give this last answer? In man, at the critical moment, a spasm passes through the whole body, and similarly in the plant, I find that a great contractile spasm takes place. This is accompanied by an electrical spasm also. In the script of the Morograph, or Death recorder, the line that up to this point was being drawn, becomes suddenly reversed, and then ends. This is the last answer of the plant.

These are mute companions, silently, growing beside our door, have now told us the tale of their life-tremulousness and their death spasm, in script that is as inarticulate as they. May it not be said that this their story has a pathos of its own, beyond any that the poets have conceived?

PROF. J. C. BOSE AT MAYAVATI

MARVELS OF PLANT LIFE

On the 8th June 1912, Dr J. C. Bose, who had gone to Advaita Ashrama, Mayavati, on a holiday trip, gave an illuminating discourse on the marvels of plant life.

He began by stating that a stimulus takes a certain time before it gets a response. This stimulus may be of different forms, *e.g.*, it may be a sound stimulus, a light stimulus, an electric stimulus, and so on. The feebler the stimulus, the greater is the time it takes to elicit the response. For instance if one is called by a distant voice, one doubts whether he has been called at all, but in the case of a piercing scream, he starts up at once.

Now, the difficulty is that when the stimulus, the blow, is so strong as to get an instantaneous response, how is one to measure this infinitesimal time between the blow and the response? And this must be done absolutely free from any personal interference, so as to ensure correct results.

[Pg_103] Dr. Bose here described how after deep thought and careful experiments and researches of several years he invented and manufactured a highly sensitive instrument which could automatically record the "response time" of a plant even to one thousandth part of a second. And in order to convey a graphic idea of the principles under which it worked, he had even made by means of a few simple things a crude form of his instrument, which helped the audience to form a clear idea of how a shock given to a plant which was experimented upon, would be recorded automatically by the apparatus by means of dots on its writing pad, and also how to ascertain the exact time each plant took to respond to the stimulus received. Thus the plant now records its own history unerringly by its own hand as it were. And that the *same* results are obtained each time the experiment is repeated under similar conditions, shows that this recording of the response time is a scientific phenomenon.

As an example of the similarities of reactions in plant and animal, Prof. Bose described the rhythmic activities of certain plants, in which automatic pulsations are maintained as in the animal heart. This phenomenon is exemplified by the Telegraph plant, which grows wild in the Gangetic plane; its Indian name is *Bon charal* or 'forest churl', the popular belief being that it dances to the clapping [Pg_104] of the hand. There is no foundation however for this belief. It is a papilionaceous plant with trifoliate leaves, of which the terminal leaflet is large, and the two lateral, very small. Each of these is inserted on the petiole by means of pulvinule. The lateral leaflets are seen to execute pulsating movements which are apparently uncaused, and are not unlike the rhythmic movement of the heart to which we shall see later that their resemblance is more than superficial.

In the intact plant, under favourable conditions, these movements are easily observed to take place more or less continuously; but there are times when they come to a standstill. For this reason and because of the fact that a large plant cannot easily be manipulated as a whole and subjected to various changing conditions which the purpose of the investigation demands, it is desirable, if possible, to experiment with the detached petiole, carrying the pulsating leaflet. The required amputation however may be followed by arrest of the pulsating movements. But, as in the case of the isolated heart in a state of standstill, Dr. Bose found that the movement of the leaflet can be renewed, in the detached specimen, by the application of the internal hydrostatic pressure. Under these conditions, the rhythmic pulsations are easily maintained uniform for several hours. This is a great advantage, [Pg_105] in as much as in the undetached specimen, the pulsations are not usually found to be so regular as they now become. So small a specimen, again, can easily be subjected to changing experimental conditions, such as the variation of internal hydrostatic pressure and temperature, application of different drugs, vapours and gases.

Under varying conditions the same plant has been observed to take different response times, as for instance, less in heat than in cold, less in summer than in winter, less in the morning than in the evening, and so forth. Again, different plants have different response times.

It is a remarkable fact that the mimosa is ten times as sensitive as a frog in giving the response. And the native idea that plants are of a lower order than animal life will cost many a sad disappointment.

In the course of his lecture Dr. Bose spoke of some of his startling discoveries recently made.... The lecturer gave quite a spiritual turn to his discourse as he finished it with the remark that, as it has been the earnest endeavour of scientists to minimise material friction in order to get the best results, so in our human concerns, it should be our best aim to minimise friction,—which is, Ignorance.

—*Modern Review*, Vol. XII, pages 314-315.

PLANT AUTOGRAPHS

HOW PLANTS CAN RECORD THEIR OWN STORY

Under the presidency of His Excellency Lord Carmichael, Prof. J. C. Bose delivered on Friday, the 17th January 1913 an interesting address on his recent researches at the Physical Laboratory of the Presidency College, Calcutta, his subject being "Plant Autographs."

Professor Bose has been long engaged in researches on the "Irritability of Plants," with results of great interest. These results have been made possible by the invention of a series of instruments of extraordinary precision and delicacy. Some of Professor Bose's instruments measure and record a thousandth of a second. Invisible movements in plants, hitherto beyond human scrutiny, have been brought within the range of immediate perception through the wonderful devices shown by the lecturer's demonstration of same on the screen.

Among those present were:—Sir William and Lady Duke, the Maharaja of Nashipur, Sir Gurudas [Pg_107] Bannerjee, Sir Chundra Madhab Ghose, Sir Lawrence and Lady Jenkins, Sir Richard Harington, Hon. Mr. P. C. Lyon, Mr. Justice Holmwood, Mr. Justice Chaudhuri, Hon. Mr. S. L. Maddox, Maharaja of Cossimbazar, Hon. Dr. Kuchler, Mr. Bhupendra Nath Basu, Hon. Mr. E. W. Collin, Mr. W. Graham, Mr. Fraser Blair, Hon. Mr. B. Chuckerbutty, Hon. Mr. J. G. Apcar, Hon. Mr. B. C. Mitter, Hon. Rai Radha Charan Pal Bahadur, Hon. Dr. D. P. Sarbadhikari, Mr. and Mrs. Williams, Mr. L. P. E. Pugh, Mr. Lanford James, Dr. P. K. Roy, Khan Bahadur Moulvie Mahomed Yusuf, Rai Bahadur Dr. Chunilal Bose, Mr. W. J. Simmons, Mr. and Mrs. J. H. Hechle, Principal H. R. James and Mrs. James, Mr. T. J. Waite, Dr. P. C. Roy and Rai P. N. Mukherji Bahadur.

His Excellency, as President, called upon Dr. Bose to deliver his lecture.

Professor Bose commenced with a reference to the claims made by those who profess to discriminate character by handwriting. As to the authenticity of such claims, scepticism was permissible; but there was no doubt that one's handwriting might be modified profoundly by conditions, physical and mental. There still existed, at Hatfield House, documents which contained the signature of the historical Guy Fawkes. A photograph projected on the screen showed a sinister variation in those [Pg_108] signatures. The crabbed and distorted characters of the last words which Guy Fawkes wrote on earth told their own tale of that fateful night. Such was the tale that might be unfolded by the lines and curves of a human autograph. Could plants be made similarly to write their own autographs revealing their hidden story? Storm and sunshine, the warmth of summer and the frost of winter, drought and rain, would come and go about the plants. What subtle impress did they leave behind? How were the invisible, internal changes to be made externally visible?

AUTOMATIC RECORDERS

The lecturer had succeeded in devising experimental methods and apparatus by which the plant was made to give an answering signal, which was then automatically recorded into an intelligible script. The results of the new investigations were so novel that Professor Bose spent several years in perfecting automatic instruments which completely eliminated all personal equations. The plant attached to the recording apparatus was automatically excited by a stimulus absolutely constant, making its own responsive records, going through its period of recovery, and embarking on the same cycle over again without assistance at any point [Pg_109] from the observer. The most sensitive organ for perception of a stimulus was the human tongue. An average European could by his tongue detect an electrical current as feeble as six micro-amperes, a micro-ampere being a millionth part of a unit of electrical current. Professor Bose found that his Hindu peoples could detect a much feebler current, namely, 1.5 micro-amperes. It was an open question whether such a high excitability of the tongue was to be claimed as a distinct advantage. But the fact might explain the eminence of his countrymen in forensic domains! (Laughter.) The plant, when tested, was found to be ten times more sensitive than a human being.

EFFECT OF FOOD AND DRUGS

It was shown that when the plant had a surfeit of drink, it became excessively lethargic and irresponsive. By extracting fluid from the gorged plant, its motor activity was at once re-established. Under alcohol its responsive script became ludicrously unsteady. A scientific superstition existed regarding carbonic acid as being good for a plant. But Professor Bose's experiments showed distinctly that the gas would suffocate the plant as readily as it did the animal. Only in the presence of sunlight could the effect be modified by secondary reaction.

AUTOMATISM [Pg_110] AND GROWTH

It was impossible in a limited space, said Professor Bose, to do more than mention the numerous other remarkable experiments which riveted the attention of the audience. By means of apparatus specially devised, pulsative plants were made to record their rhythmic throbbings. It was shown that the pulse beats of the plants were affected by the action of various drugs, and divers stimuli, in a manner similar to that of the animal heart. Perhaps the most weird experience was to watch the death-struggle of a plant under the action of poison. Turning from death to its antithesis life and growth, the audience were shown how the latter was made visible by means of the appliances invented by Professor Bose. The infinitesimal growth of a plant became highly magnified in the experiment.

RESEARCHES AT PRESIDENCY COLLEGE

When the lecturer commenced his investigations, original research in India was regarded as an impossibility. No proper laboratory existed, nor was there any scientific manufactory for the construction of a special apparatus. In spite of these difficulties it had been a matter of gratification to the lecturer that the various investigations already carried out at the Presidency College had done something for the advancement of knowledge. The [Pg_111] delicate instruments seen in operation at the lecture, which had been regarded with admiration by many distinguished scientific men in the West, were all constructed at the College workshops by Indian mechanics.

It was also with pride that the lecturer referred to the co-operation of his pupils and assistants, through whose help the extensive works, requiring ceaseless labour by day and night, had been accomplished. Doubt had been cast on the capacity of Indian students in the field of science. From his personal experience Professor Bose bore testimony to their special fitness in this respect. An intellectual hunger had been created by the spread of education. An Indian student demanded something absorbing to think about and to give scope for his latent energies. If this could be done, he would betake himself ardently to research into Nature, which could never end. There was room for such toilers who by incessant work would extend the bounds of human knowledge.

FROM PLANT TO ANIMAL LIFE

Before concluding the lecturer dwelt on the fact that all the varied and complex responses of the animal had been foreshadowed in the plant. The phenomena of life in the plant were thus not so remote as had been hitherto supposed. The plant [Pg_112] world, like the animal, was a thrill and a throb with responsiveness to all the stimuli which fell upon it. Thus, community throughout the great ocean of life, in all its different forms, outweighed apparent dissimilarity. Diversity was swallowed up in unity.

—*Amrita Bazar Patrika*, 20-1-1913.

INVISIBLE [Pg_113] LIGHT

A most instructive and interesting lecture was delivered on Thursday, the 30th Jaunuary, 1913, at the Calcutta University Institute Hall, by Dr. J. C. Bose, on the above subject. It was illustrated with experiments and in spite of the technical nature of the subject, the manner of treatment made the discourse extremely palatable and easy of apprehension to the lay understanding and intelligence. The truths of science could seldom be exposed so light-heartedly and in language leavened with balmy humour. The lecture was very largely attended by ladies and gentlemen, European and Indian, representing the light and leading of the city. The chair was taken by Mr. W. R. Gourlay. Amongst those present we noticed the Hon. Mr. Ramsay McDonald, Mr. Justice Harington, Mr. Justice Chaudhuri, Hon'ble Mr. Gokhale, Hon'ble Mr. Lyon, Hon'ble Mr. D. N. Sarvadhikari, Sir Gurudas Banerji, Hon'ble Mr. Apcar and Dr. Chuni Lal Bose Rai Bahadur.

The Chairman, in a few well chosen words introduced the lecturer.

[Pg_114] Professor Bose in going to deliver his highly interesting lecture first showed how on account of the imperfection of our senses we fail to detect various forces which play around us. We are not only deaf, but practically blind. While we perceive eleven octaves of sound, we can see only a single octave of other vibration which is called light. In order to detect the invisible light a special detector has to be devised. Prof. Bose showed his artificial retina previously exhibited at the Royal Institution which not only detected luminous radiation but also invisible lights in the intra red and ultra violet regions. In the course of his remarks illustrating the nature of electric or Hertzian waves, which gave rise to the invisible radiation he proceeded to enumerate some of the conditions necessary for experimenting with them, and to describe the apparatus he had invented for the purpose. Hertz had used waves which were about 10 metres in length. It was impossible to attempt any quantitative measurement of their optical properties on account of large waves curling round corners. The lecturer had succeeded in producing the shortest waves, with frequency of 50,000 millions of vibrations per second, the particular invisible radiation being only thirteen octaves below visible light. His generator produced the small sharp beam which alone could be employed for quantitative [Pg_115] measurements. By means of this apparatus experiments on electric radiation could be carried on with as much certainty as could experiments with ordinary light. Prof. Bose then performed experiments illustrative of the properties possessed in common by light waves and electric waves. He exhibited the power of selective absorption to electric rays displayed by many substances pointing out that while water stopped them, pitch, coal tar, and others were quite transparent to them. He showed how the rays were reflected by mirrors, obeying the same laws as light. The hand of the experimenter was found to be a good reflector, the rays rebounding after impact. Electric rays also undergo refraction and he described an ingenious method he had devised by which the index of refraction of numerous opaque substances could be obtained with the highest exactitude. In conclusion he gave an account of his discovery of the polarisation of electric rays by crystals. He showed that these polarised the electric rays just as they did ordinary light. He further proved that substances under pressure and strain could produce double refraction in them, as did glass under the same conditions in light. Tourmaline was useless for electric rays; but a lock of human hair was extraordinarily efficient. According to this theoretical prediction, an ordinary book was shown to exhibit [Pg_116] selective absorption in a striking manner. Thus while the Calcutta University Calendar was, usually, very opaque, it became quite transparent when held in a particular direction as regards the impinging ray.

Mr. Gourlay observed that the lecture opened out to himself, as well as to other vistas, which they had never dreamt of before.

—*Amrita Bazar Patrika*, 31-1-1913.

PROFESSOR J. C. BOSE AT LAHORE

LECTURE ON ELECTRIC RADIATION

A crowded assembly met at the University Hall, on the 22nd February, 1913, to hear the first of Prof. Bose's discourses before the University of Lahore.

Dr. Bose opened his address by alluding to the historic journey of Jivaka, who afterwards became the physician of Buddha, making his way from Bengal to the University of Taxila, in quest of knowledge. Twenty-five centuries had gone by and there was before them another pilgrim who had journeyed the same distance to bring, as an offering what he had gathered in the domain of knowledge.

The lecturer called attention to the fact that knowledge was never the exclusive possession of any particular race nor did it ever recognise geograpahical geographical limitations. The whole world was interdependent, and a constant interchange of thought had been carried on throughout the ages enriching the common heritage of mankind. Hellenistic Greeks and Eastern Aryans had met here in Taxila to exchange the best each had to offer. [Pg_118] After many centuries the East and West had met once more, and it would be the test of the real greatness of the two civilisations that both should be finer and better for the shock of contact. The apparent dormancy of intellectual life in India had been only a temporary phase. Just like the oscillations of the seasons found the globe, great pulsations of intellectual activity pass over the different peoples of the earth.

With the coming of the spring the dormant life springs forth; similarly the life that India conserves, by inheritance, culture and temperament, was only latent and was again ready to spring forth into the blossom and fruit of knowledge. Although science was neither of the East nor of the West, but international in its universality, certain aspect of it gained richness of colour by reason of their place of origin. India, perhaps through its habit of synthesis, was apt to realise instinctively the idea of unity and to see in the phenomenal world an universe instead of a multiverse. It was this tendency, the lecturer thought, which had led Indian physicist, like himself, when studying the effect of forces on matter to find boundary lines vanishing, and to see points of contact emerge between the realms of the living and non-living. In taking up the subject of the evening's discourse on electric radiation of Hertzian waves, the lecturer explained [Pg_119] the constitution of the apparatus which he had devised for an exhaustive study of the properties of electric waves. His apparatus permitted experiments with the electric rays to be carried on with as much certainty as experiments with ordinary light, and he demonstrated the identity of electric radiation and light. The electric rays are reflected from plane and curved mirrors in the same way and subject to the same laws. Electric rays, like rays of light are refracted. Like race of light too, electric waves can be selectively stopped by various substances, which are "electrically" coloured. Water which is a conductor of electricity stops the electric ray; where as liquid air which is a non-conductor is quite transparent to the rays.

Finally Professor Bose explained his discovery of Polarisation of these rays by various crystals. Tourmaline, which was a good polariser for ordinary light, was not so effective. The lecturer discovered that the crystal Nemalite possessed the power of polarising the electric rays in the most perfect manner. Professor Bose also explained how the internal constitution of an opaque mass was revealed by the help of light which was itself invisible.

The lecturer concluded his discourse by drawing attention to the limitations of human perception. Man's power of hearing was confirmed to eleven octaves of sound notes. In the case of vision the [Pg_120] limitation was far more serious, his power of sight extending only through a single octave of those ether waves which constituted light. These ether vibrations of various frequencies could be maintained by electrical means. By pressing the stop button of the apparatus which was exhibited, ether vibrations, 50,000 millions per second, were produced. A second stop gave rise to a different vibration. Let his audience imagine a large electric organ provided with an infinite number of stops, each stop giving rise to a particular ether note. Let the lowest stop produce one vibration a second. They should then get a gigantic wave of 186,000 miles long. Let the next stop give rise to two vibrations in a second, and let each succeeding stop produce higher and higher notes. Let them imagine an unseen hand pressing the different stops in rapid succession, producing higher and higher notes. The ether note would thus rise in frequency from one vibration in a second, to tens, to hundreds, to thousands, to hundreds of thousands, to millions, to millions of millions! While the ethereal sea in which they were all immersed were being thus agitated by these multitudinous waves, they would remain entirely unaffected, for they possessed no organs of perception, to respond to these waves.

As the ether note rose still higher in pitch, they would for a brief moment perceive a sensation of [Pg_121] warmth. This would be the case when the ether vibration reached a frequency of several billions of times in a second. As the note rose still higher, their eyes would begin to be affected, a red glimmer of light being the first to make its appearance. From this point the few visible colours would be comprised within a single octave of vibration—from 400 to 800 billions in one second. As the frequency of vibration rose still higher their organs of perception would fail them completely; a great gap in their consciousness would obliterate the rest. The brief flash of light would be succeeded by unbroken darkness. How circumscribed was their knowledge? In reality they stood in the midst of a luminous ocean almost blind! The little they could see was as nothing compared to the vastness of that which they could not. But it may be said that, out of the very imperfection of his senses, man has been able, in science, to build for himself a raft of thought by which to make daring adventure on the great seas of the unknown.

—*Amrita Bazar Patrika*, 24-2-1913.

DR. BOSE IN LAHORE

PLANT RESPONSE

In his third lecture delivered, on the 25th February 1913, at the Punjab University Hall, Dr. Bose of Calcutta dealt with "Plant Response." He said: —

In strong contrast to the energetic animal, with its various reflex movements and pulsating organs, stands the plant, in its apparent placidity and immobility. Yet that same environment which with its changing influences affects the animal is playing upon it also. Storm and sunshine, the warmth of summer and the frost of winter, drought and rain, all these come and go about it. What coercion do they exercise upon it? What subtle impress do they leave behind? These internal changes are entirely beyond our visual scrutiny. Is it possible in any way to have these revealed to us? Dr. Bose had shown the possibility of this by detecting and measuring the actual response of the organism to a questioning shock. In an excitable condition the feeblest stimulus should evoke in the plant an extraordinarily large reply in a depressed state even [Pg_123] a strong stimulus would only call forth a feeble response; and lastly, when death overcome life, there would be an abrupt end of the power to answer to all. By the invention of different types of apparatus, the lecturer had succeeded in making the plant itself write an answering script to a testing stimulus. Scripts could also be obtained of the plant's spontaneous movements; and a recording arm demarcated the line of life from that of death.

In taking the self-made records made by the plant it was found that after the prolonged inactivity of a cold night the plant was apt to be lethargic, and its first answers indistinct. But as blow after blow was delivered, the lethargy passed off, and the replies became stronger and stronger. After the fatigue of the day, the state of things was reversed. The plant became very lethargic after excessive absorption of food; but the normal activity might be restored by artificial removal of the excess. The effect of alcohol and of various narcotics were clearly followed in the modification of the automatic record made by the plant.

A prevailing scientific error had overcome in life, there would be an abrupt end regarding a certain class of plants to be alone sensitive. The lecturer showed by certain remarkable experiments that all plants and all organs of plants were sensitive.

In certain animal tissues, a very curious phenomenon [Pg_124] was observed. In man and other animals there were tissues which beat spontaneously. As long as life lasted, so long did the heart continue to pulsate. There could be no effect without a cause. How then was it that these pulsations became spontaneous? To this query, no satisfactory answer had been forthcoming. Similar spontaneous movements were also observable in plant tissues, and by their investigation the secret of automatism in the animal world became unravelled. The existence of these spontaneous movements could easily be demonstrated by means of the Indian "Bon Charal", the telegraph plant, whose small leaflets danced continuously up and down. The popular belief that they danced in response to the clappings of the hand was quite erroneous. From the readings of the scripts made by this plant, the lecturer was in a position to state that the automatic movements of both plants and animals were guided by laws which were identical. Thus in the rhythmic tissues of the plant and the animal the pulsation frequency was increased under the action of warmth and lessened under cold, increased frequency being attended by diminution of amplitude, and *"vice versa"*. Under ether, there was a temporary arrest, revival being possible when the vapour was blown off. More fatal was the effect of chloroform. The most extraordinary parallelism, however, lay in the [Pg_125] fact that those poisons which arrested the beat of the heart in a particular way arrested the plant pulsation in a corresponding manner. The lecturer had succeeded in reviving a leaflet poisoned by the application of one with a dose of counteracting poison.

A time came when after one answer to a supreme shock there was a sudden end of the plant's power to give any response. This supreme shock was the shock of death. Even in this crisis, there was no immediate change in the placid appearance of the plant. In man at the critical moment, a spasm passed through the whole body, and similarly in the plant the lecturer had discovered that a great contractile spasm took place. This was accompanied by an electrical spasm also. In the script of the death recorder the line that up to this point was being drawn became suddenly reversed, and then ended. This was the last answer of the plants.

Thus the responsiveness of the plant world was one. There was no difference of any kind between sunshine plants, and those which had hitherto been regarded as insensitive or ordinary. It had also been shown that all the varied and complex responses of the animal were foreshadowed in the plant. An impressive spectacle was thus revealed of that vast unity in which all living organisms, from the simplest plant to the highest animal, were linked together and made one.

—*Amrita Bazar Patrika*, 5-3-1913.

EVIDENCE BEFORE THE PUBLIC SERVICES COMMISSION

The following is the evidence given by Dr. J. C. Bose, C. S. I., C. I. E., Professor of Physics, Presidency College, Calcutta, on the 18th December, 1913, before the Royal Commission on the Public Services in India, presided over by Lord Islington, and published, in the Minutes of Evidence relating to the Education Department, at pages 135 to 137, in volume XX, Appendix to the Report of the Commissioners:

WRITTEN STATEMENT RELATING TO THE EDUCATION DEPARTMENT

83, 627 (I) *Method of recruitment.*—The first question on which I have been asked to give my opinion is as regards the method of recruitment. I think that a high standard of scholarship should be the only qualification insisted on. Graduates of well-known Universities, distinguished for a particular line of study, should be given the preference. I think the prospects of the Indian Educational Service are [Pg_127] sufficiently high to attract the very best material. In colonial Universities they manage to get very distinguished men without any extravagantly high pay. Possibly the present departmental method of election does not admit of sufficiently wide publicity of notice to attract the best candidates.

83, 628 (II) *System of training and probation.*—As regards probation and training, Educational officers should first win a reputation as good teachers before the appointment is confirmed as they are transferred to important colleges.

83, 629 (IV) *Conditions of Salary.*—As regards conditions of Salary, the pay should be moderately high, but not extravagant, and settled once for all under some simple and well-defined rules. It is not only very humiliating but degrading to a true scholar to be scrambling for money. The difference between the pay of the higher and lower services should be minimised.

83, 630 (VI) *Conditions of pension.* — With reference to pension, I think it is very unfair that more favourable terms are offered, when the pensioner elects to retire in England.

83, 631 (VII) *Such limitations as exist in the employment of non-Europeans.* — Passing on to the question of limitations that exist in the employment of Indians in the higher service, I should like to give expression to an injustice which is very keenly felt. It is [Pg_128] unfortunate that Indian graduates of European Universities who have distinguished themselves in a remarkable manner do not for one reason or other find facilities for entering the higher Educational Service.

As teachers and workers it is an incontestable fact that Indian officers have distinguished themselves very highly, and anything which discriminates between Europeans and Indians in the way of pay and prospects is most undesirable. A sense of injustice is ill-calculated to bring about that harmony which is so necessary among all the members of an educational institution, professors and students alike.

83, 632 (VIII) *Relations of the service with the Indian Civil Service and with other services.* — As regards the relations with the Indian Civil Service, I am under the impression that they are somewhat strained, but of this I have no personal experience.

83, 633 (IX) *Other points.* — I have endeavoured to give my opinion on the definite questions which have been asked. There is another aspect of educational work in India which I think of the highest importance, though I am not exactly sure whether it falls within the terms of reference to the Royal Commission. I think that all the machinery to improve the higher education in India would be altogether ineffectual unless India enters the world [Pg_129] movement for the advancement of knowledge. And for this it is absolutely necessary to touch the imagination of the people so as to rouse them to give their best energies to the work of research and discovery, in which all the nations of the world are now engaged. To aim at anything less will only end in a lifeless and mechanical system from which the soul of reality has passed away. On this subject I could have said much, but I will confine myself to one point which I think at the present juncture to be of importance. The Government of Bengal has been foremost in a tentative way in encouraging research. What is necessary is the extension and continuity of this enlightened policy.

83, 634. *Supplementary Note.* — I would like to add a few remarks to make the meaning of paragraphs 83, 627 and 83, 631 in my note more explicit.

At the present recruitment in the Indian Educational Service is made in England and is practically confined to Englishmen. Such racial preference is in my opinion, prejudicial to the interest of education. The best man available, English or Indian should be selected impartially, and high scholarship should be the only test.

It has been said that the present standard of Indian Universities is not as high as that of British Universities, and that the work done by the former is more like that of a sixth form of public schools in [Pg_130] England. It is therefore urged that what is required for an Educational officer is the capacity to manage classes rather than high scholarship. I do not agree with these views: (1) there are Universities in Great Britain whose standards are not higher than ours; I do not think that the Pass Degree even of Oxford or Cambridge is higher than the corresponding degree here; (2) the standard of the Indian Universities is being steadily raised; (3) the standard will depend upon what the men entrusted with Educational work will make it. For these reasons it is necessary that the level of scholarship represented by the Indian Educational Service should be maintained very high.

In paragraph 83,631 I have stated that even these Indians who have distinguished themselves in European Universities have little chance of entering the higher Educational Service. I should like to add that these highly qualified Indians need only opportunities to render service which would greatly advance the cause of higher education. As regards graduates of Indian Universities, I have known men among them whose works have been highly appreciated. If promising Indian graduates are given the opportunity of visiting foreign Universities, I have no doubt that they would stand comparison with the best recruits that can be obtained from the West.

DR. [Pg_131] J. C. BOSE CALLED AND EXAMINED

83,635. (Chairman). The witness favoured an arrangement by which Indians would enter the higher ranks of the service, either through the Provincial Service or by direct recruitment in India. The latter class of officers, after completing their education in India, should ordinarily go to Europe with a view to widening their experience. By this he did not wish to decry the training given in the Indian Universities, which produced some of the very best men, and he would not make the rule absolute. It was not necessary for men of exceptional ability to go to England in order to occupy a high chair. Unfortunately, on account of there being no openings for men of genius in the Educational Service, distinguished men were driven to the profession of Law. In the present condition of India a larger number of distinguished men were needed to give their lives to the education of the people.

83,636. The witness himself had spent part of his career in Europe, and looking back he could say that this had been of great profit to him, not so much on account of the training he got, as by being brought into personal contact with eminent men whose influence extorted his admiration, and create in him a feeling of emulation. In this way he owed a great deal to Lord Rayleigh under whom he [Pg_132] worked, but he did not see why that advantage should not eventually be secured by Indians in India under an Indian Lord Rayleigh.

83,637. There should be only one Educational Service, but men who were distinguished in any subject should not start from its very lowest rung but should be placed somewhere in the middle of it.

83,638. There were men in the Provincial Service who were very distinguished; it was all a question of genius. The Educational Service ought to be regarded not as a profession, but as a calling. Some men were born to be teachers. It was not a question of race, of course; in order to have an efficient educational system, there must be an efficient organisation, but this should not be allowed to become fossilised, and thus stand in the way of healthy growth.

83,639. In the Presidency College a young man fresh from an English university was at once appointed a Professor regardless of his lack of experience, whereas an Indian who passed in highest examination with honours in India was appointed as an Assistant Professor. This grounding often made him more efficient as a teacher than the Professor recruited from England. There were now several Professors in the college, in the Provincial Service, who were highly qualified, and who lectured to the highest classes with very great success.

[Pg_133] 83,640. In the Physics Department he had under his direction several Assistants who were so well qualified that they were allowed to give lectures to several classes. These Assistants, after their experience at the Presidency College, would be best fitted to become Professors in the mofussil at Colleges. He would like to see them promoted to the higher service after they had had experience. But before he gave them the highest positions, he would make it compulsory for them to go to Europe.

83,641. A proportion of Europeans in the service was needed, but only as experts and not as ordinary teachers. Only the very best men should be obtained from Europe, and for exceptional cases. The general educational work should be done entirely by Indians, who understood the difficulties of the country much better than any outsider.

83,642. He advocated the direct recruitment of Indians in India by the local government in consultation with the Secretary of State, rather than by the Secretary of State alone. Indians were under a great difficulty, in that they could not remain indefinitely in England after taking their degrees and being away from the place of recruitment their claims were overlooked.

83,643. There was no reason why a European should be paid a higher rate of salary than an Indian on account of the distance he came. An [Pg_134] Indian felt a sense of inferiority if a difference was made as regards pay. The very slight saving which government made by differentiating between the two did not compensate for the feeling of wrong done. This feeling would remain even if the pay was the same, but an additional grant in the shape of a foreign service allowance was made to Europeans. All workers in the field of education should feel a sense of solidarity, because they were all serving one great cause, namely, education.

83,644. The term "professor", as at present used in India, was undoubtedly a comprehensive one, but it was equally comprehensive in the West.

83,645. (Sir Murray Hammick). The witness did not wish to recruit definite proportions of the service in England and in India respectively. He would for various reasons prefer a large number of Indians engaged in education.

83,646. Even in Calcutta he would not make any difference between the pay of the Indian and the pay of the European.

83,647. (Sir Valentine Chirol). The witness attached great value to the influence of the teacher upon the student in the earlier stages of his education, and it was in these stages that that influence could best be exercised. At the same time he desired to limit the appointment of non-Indians to men of very great distinction.

[Pg_135] 83,648. If a foreign professor would not come and serve in India for the same remuneration as he obtained in his own country, the witness would certainly not force him to come.

83,649. (Mr. Abdur Rahim). Recruitment for the Educational Service should be made in the first place in India, if suitable men were available; but if not then he would allow the best outsiders to be brought in. In the present state of the country it would be very easy to fill up many of the chairs by selecting the best men in India.

83,650. The aim of the universities should be to promote two classes of work—first, research; and secondly, an all-round sound education. Men of different types would be required for these two duties.

83,651. (Mr. Madge). Any idea that the educational system of India was so far inferior to that of England, that Indians, who had made their mark, had done so, not because of the educational system of the country, but in spite of it, was quite unfounded. The standard of education prevailing in India was quite up to the mark of several British universities. It was as true of any other country in the world as of India that education was valued as a means for passing examinations, and not only for itself, and there was no more cramming in India than elsewhere.

[Pg_136] 83,652. The West certainly brought to the East a modern spirit, which was very valuable, but it would be dearly purchased by the loss of an honorable career for competent Indians in their own country.

83,653. The educational system in India had in the past been too mechanical, but a turn for the better was now taking place and the universities were recognising the importance of research work, and were willing to give their highest degrees to encourage it.

83,654. (Mr. Macdonald). The witness did not think it was necessary to have a non-Indian element in the service in order to stiffen it up, but he accepted the principle that there should be a certain small proportion of non-Indians.

83,655. The title of professor at a college or University should carry with it dignity and honour, and ought not to be so freely used as at present. All he asked was that it should not be abolished at the expense of such Indians as were doing as good work as their European colleagues.

83,656. If the Calcutta university continued to develop its teaching side, there would be no objection to recruiting University Professors from aided colleges. This would have certain advantages.

83,657. (Mr. Fisher). The witness desired to secure for India Europeans who had European [Pg_137] reputations in their different branches of study. If it was necessary to go outside India or England to procure good men, he would prefer to go to Germany. This was the practice in America where they were annexing all the great intellects of Europe.

83,658. The witness would like to see India entering the world movement in the advance and march of knowledge. It was of the highest importance that there should be an intellectual atmosphere in India. It would be of advantage if there were many Indians in the Educational Service. For they came more in contact with the people, and influenced their intellectual activity. Besides, on retirement they would live in India and their life experience would be at their countrymen's service.

83,659. There was very little in the complaint made in certain quarters that the work of the Professors in the colleges in India was hampered by the Government regulations as to curricula. A good teacher was not troubled by such matters.

83,660. (Mr. Sly). There was no scope for the employment of non-Indians in the high schools as apart from the colleges. It was in the professorial line that more help from the West was required.

83,661. (Mr. Gokhale). The witness knew of three instances in which the colonies had secured distinguished men on salaries which were lower than these given to officers of the Indian Educational [Pg_138] Service. One was at Toronto, another was in New Zealand and the third at Yale university. The salaries on the two latter cases were #600 and #500 a year. The same held good as regards Japan. The facts there had been stated in a Government of India publication as follows: "Subsequent to 1895 there were 67 Professors recruited in Europe and America, of those, 20 came from Germany, 16 from England and 16 from the United States. The average pay was #384. In the highest Imperial University the average pay is #684. As soon as Japanese could be found to do the work, even tolerably well, the foreigner was dropped."

83,662. When the witness first started work in India, he found that there was no physical laboratory, or any grant made for a practical experimental course. He had to construct instruments with the help of local mechanics, whom he had to train. All this took him ten years. He then undertook original investigation at his own expense. The Royal Society became specially interested in his work and desired to give him a Parliamentary grant for its continuation. It was after this that the Government of Bengal came forward and offered him facilities for research.

83,663. In the Educational Service he would take men of achievement from anywhere; but men of promise he would take from his own country.

[Pg_139] 83,664. (Mr. Chaubal). He did not know whether the salaries he had mentioned as having been paid in Japan, New Zealand and Yale were on an incremental scale or not.

83,665. There was a difference of kind between the way in which students were taught in schools and the way in which they were taught in colleges. He did not agree with the witnesses who had said that during the first year or two years at college the instruction given was similar to that given in a school. It was very difficult to disprove or to prove such statements. There would be no advantage in keeping boys to a school course up the intermediate standard and making the colleges deal with only those students who had passed the intermediate examination.

83,666. (Sir Theodore Morison). There should be one scale of pay for all persons in the higher educational department. The rate of salary, Rs. 200 rising to Rs. 1,500 per month, was suitable, subject to the proviso that the man of great distinction, instead of beginning at the lowest rate of pay, should start some where in the middle of the list, say, at Rs. 400 or Rs. 500. He would make no reference in regard to Europeans or Indians in that respect. In effect this no doubt amounted to making Indians eligible for higher educational posts both by direct recruitment and by promotion.

[Pg_140] 83,667. He would not favour the handing over of all the Government institutions in Bengal to private agencies; there must be one or two Government colleges in order to keep up the standard. He should be sorry to see the Government dissociating itself from one of its primary duties, which was education.

83,668. Privately managed Colleges paid less in salary than the Government Colleges. They paid about the same as was given in the Provincial Service, and they obtained fairly good men. It would not be right for a great Government to grant a minimum pay to Indian Professors and an extravagantly high pay to their European colleagues, for doing the same kind of work.

83,669. At the Presidency College the facilities for scientific work were now greater than in many institutions in England. India was now becoming a great country for Biological research. Again, the Physical and Chemical Laboratories at the Presidency College were finer than many in England. If young men of science in England thought they obtained better opportunities in pursuing their subjects in New Zealand and Toronto than in India, the India office ought to remove that impression at once.

83,670. (Lord Ronaldshay). When an Indian graduate under the witnesses' scheme was appointed direct to the higher service in India he would [Pg_141] not compel him to go to England for a period of training. The person who would be appointed in India directly from the Indian Universities would have to have previously served with distinction in subordinate positions; a visit to Europe would be an advantage but not absolutely necessary.

83,671. (Mr. Biss). The cost of living in Calcutta to an Indian Professor or Lecturer would all depend as the style in which he lived. In each service there is always a standard of living to which every member is expected to conform. An Indian Professor had to go to Europe from time to time to keep himself in touch with the developments of his subject. An Indian officer had to support a large number of relations. The question of a man's private expenses should not be raised in fixing his pay. One might as well inquire whether the candidate for admission to the service was a bachelor or married, or as to how many children he had. He had known Europeans who had led a simple life, and had been all the better for it.

83,672. He could not understand why men went to Japan and Canada instead of coming to India on better terms. It was a mystery to him. He thought it was either sheer ignorance or the spread of the commercial spirit.

83,673. All the students coming to his side of the University, were, as a rule, keen and [Pg_142] anxious to learn; he could not wish for better students.

83,674. (Mr. Gupta). He desired one service, because he thought it was most degrading that certain men, although they were doing the same work, should be classed in a Provincial Service, while others should be classed in an Imperial Service. The prospect of the members of the Provincial Service were not at all what they ought to be, and that was the reason why the best men were not attracted to it.

PROF. [Pg_143] J. C. BOSE AT MADURA

On his way back to Calcutta from the Fourth Scientific Deputation to the West, Prof. J. C. Bose visited Madura, 14th June 1915. The Tamil Sangam presented him with an address. In reply Dr. Bose made an important speech, in course of which he said:—

I am no longer a representative of Bengal nor have I come to a strange place, but as an Indian addressing the mighty India and her people. When we realise that unity of our destiny then a great future opens out for us.

It may be we may theorise and attribute to the plants all the characteristics of the animals; but that will be merely theory: there will be no proof. There are certain classes of people who think that plants are utterly unlike animals and some hold that they are like animals. The mere theory is absolutely worthless in order to find out the truth. We have to find by investigation, by means of researches, by means of proofs, that one is identical with the other. We have not only to drop all theory but we have to make the plant itself write [Pg_144] down the answers to the questions that we have to put to them. That was the great problem,—how to make the plant itself answer and write down answers to the question....

If the plants are acted on by various medicines and drugs like ourselves, then we can create an agent or a spokesman on which we can carry out all future investigations on the action of drugs. Then there is opened out a great vista for the scientific study of medicine. And let me tell you medicine is not yet an exact science. It is merely a phase of tradition. We have not been able to make medicine scientific. Now by the data of the influence of drugs on the fundamental basis of life, as is seen in the plant, we shall be able to make the science of medicine purely scientific.

In travelling all over the world, which I have done several times, I was struck by two great characteristics of different nations. One characteristic of certain nations is living for the future. All the modern nations are striving to win force and power from nature. There is another class of men who live on the glory of the past. Now, what is to be the future of our nation? Are we to live only on the glory of the past and die off from the face of the earth, to show that we are worthy descendants of the glorious past and to show by our work, by our intellect and by our service that we [Pg_145] are not a decadent nation? We have still a great and mighty future before us, a future that will justify our ancestry. In talking about ancestry, do we ever realise that the only way in which we can do honour to our past is not to boast of what our ancestors have done but to carry out in the future something as great, if not greater than they. Are we to be a living nation, to be proud of our ancestry and to try to win renown by continuous achievements? These mighty monuments that I see around me tell us what has been done till very recent times. I have travelled over some of the greatest ruins of the Universities of India. I have been to the ruins of the University of Taxilla in the farthest corner of India which attracted the people of the west and the east. I had been to the ruins of Nalanda, a University which invited all the west to gain knowledge under its intellectual fostering. I had been all there and seen them. I have come here also and want to visit Conjeevaram. But are you to foster the dead honours or to try to bring back your University in India and drag once more from the rest of the world people who would come down and derive knowledge from India? It is in that way and that way alone we can win our self-respect and make our life and the life of the nation worthy. The present era is the era of temples of learning. In order to erect temples of learning we [Pg_146] require all the offerings of our mighty people. We want to erect temples and "viharas" which are so indispensable to the study of nature and her secrets. It is a problem which appeals to every thoughtful Indian. It is by the effort of the people and by their generosity that all these mighty temples arose; and now are we to worship the dead stones or are we to erect living temples so that the knowledge that has been made in India shall be perpetuated in India? I received requests from the different Universities in America and Germany to allow students from those countries to come and learn the science that has been initiated in India. Now, is this knowledge to pass beyond our boundaries to that again in future time we may have to go to the west to get back this knowledge or are we to keep this flame of learning burning all the time?

(*Modern Review, Vol. xviii*, p. 22-23).

DR. J. C. BOSE ENTERTAINED

PARTY AT RAM MOHAN LIBRARY

On Saturday, 24th July, 1915, the members of the Ram Mohan Library and Reading room received Dr. J. C. Bose, the President of the Library in a right royal fashion, on his return to India from his Scientific Deputation to the West.

There was a large and influential gathering, and the spacious hall was tastefully decorated.

Dr. J. C. Bose arrived at 6:15 p.m. and was received at the gate by Mr. D. N. Pal, Secretary. Dr. Bose then went round the hall accompanied by the members of the Executive Committee while the Bharati Musical Association played excellent Jaltaranga Orchestra.

Babu Bhupendra Nath Bose, Vice-President of the Library, made a brilliant speech welcoming Dr. Bose and detailing the great services done to the country by him.

DR. BOSE'S REPLY

Dr. Bose in reply expressed his thanks for the great interest shown in different parts of this country in the success of his work. This was the fourth [Pg_148] occasion on which he had been deputed to the West by the Government of India on a scientific mission, and the success that has attended his visit to foreign countries has exceeded all his expectations. In Vienna, in Paris, in Oxford, Cambridge and London, in Harvard, Washington, Chicago and Columbia, in Tokio and in many other places his work has uniformly been received with high appreciation. In spite of the fact that his researches called into question some of the existing theories, his results have notwithstanding received the fullest acceptance. This was due to a great extent to the convincing character of the demonstration afforded by the very delicate instruments he had been able to invent and which worked under extremely difficult tests with extraordinary perfection. Even the most critical savants in Vienna felt themselves constrained to make a most generous admission. In these new investigations on the border land between physics and physiology, they held that Europe has been left behind by India, to which country they would now have to come for inspiration. It has also been fully recognised that science will derive benefit when the synthetic intellectual methods of the East co-operate with the severe analytical methods of the West. These opinions have also been fully endorsed in other centres of learning and Dr. Bose had received applications from distinguished [Pg_149] Universities in Europe and America for admission of foreign post graduate scholars to be trained in his Laboratory in the new scientific methods that have been initiated in India.

RESEARCH LABORATORY FOR INDIA

This recognition that the advance of human knowledge will be incomplete without India's special contributions, must be a source of great inspiration for future workers in India. His countrymen had the keen imagination which could extort truth out of a mass of disconnected facts and the habit of meditation without allowing the mind to dissipate itself. Inspired by his visits to the ancient Universities, at Taxila, at Nalanda and at Conjevaram, Dr. Bose had the strongest confidence that India would soon see a revival of those glorious traditions. There will soon rise a Temple of Learning where the teacher cut off from worldly distractions would go on with his ceaseless pursuit after truth, and dying, hand on his work to his disciples. Nothing would seem laborious in his inquiry; never is he to lose sight of his quest, never is he to let it go obscured by any terrestrial temptation. For he is the Sanyasin spirit, and India is the only country where so far from there being a conflict between science and religion. Knowledge is regarded as religion itself. Such a misuse of science as is now unfortunately [Pg_150] in evidence in the West would be impossible here. Had the conquest of air been achieved in India, her very first impulse would be to offer worship at every temple for such a manifestation of the divinity in man.

ECONOMIC DANGER OF INDIA

One of the most interesting events in his tour round the world was his stay in Japan, where he had ample opportunity of becoming acquainted with the efforts of the people and their aspirations towards a great future. No one can help being filled with admiration for what they have achieved. In materialistic efficiency, which in a mechanical era is regarded as an index of civilisation, they have even surpassed their German teachers. A few decades ago they had no foreign shipping and no manufacture. But within an incredibly short time their magnificent lines of steamers have proved so formidable a competitor that the great American line in the Pacific will soon be compelled to stop their sailings. Their industries again, through the wise help of the State and other adventitious aids are capturing foreign markets. But far more admirable is their foresight to save their country from any embroilment with other nations with whom they want to live in peace. And they realise any predominant interest of a foreign country in their trade [Pg_151] or manufacture is sure to lead to misunderstanding and friction. Actuated by this idea they have practically excluded all foreign manufactured articles by prohibitive tariffs.

REVIVAL OF INDIAN INDUSTRIES

Is our country slow to realise the danger that threatens her by the capture of her market and the total destruction of her industries? Does she not realise that it is helpless passivity that directly provokes aggression? Has not the recent happenings in China served as an object lesson? There is, therefore, no time to be lost and the utmost effort is demanded of the Government and the people for the revival of our own industries. The various attempts that have hitherto been made have not been as successful as the necessity of the case demands. The efforts of the Government and of the people have hitherto been spasmodic and often worked at cross purposes. The Government should have an advisory body of Indian members. There should be some modification of rules as regards selection of Industrial scholars. Before being sent out to foreign countries they should be made to study the conditions of manufacture in this country and its difficulties. For a particular industry there should be a co-ordinated group of three scholars, two for the industrial and one for the commercial side. Difficulties [Pg_152] would arise in adapting foreign knowledge to Indian conditions. This can only be overcome by the devoted labour of men of originality, who have been trained in our future Research Laboratory. The Government could also materially help (i) by offering facilities for the supply of raw materials (ii) by offering expert advice (iii) by starting experimental industries. He had reason to think that the Government is full alive to the crucial importance of the subject and is determined to take every step necessary. In this matter the aims of the people and the Government are one. In facing a common danger and in co-operation there must arise mutual respect and understanding. And perhaps through the very catastrophe that is threatening the world there may grow up in India a realisation of community of interest and solidarity as between Government and people.

A CALL FOR NOBLER PATRIOTISM

A very serious danger is thus seen to be threatening the future of India, and to avert it will require the utmost effort of the people. They have not only to meet the economic crisis but also to protect the ideals of ancient Aryan civilisation from the destructive forces that are threatening it. Nothing great can be conserved except through constant effort and sacrifice. There is a danger of, regarding [Pg_153] the mechanical efficiency as the sole end of life; there is also the opposite danger of a life of dreaming, bereft of struggle and activity, degenerating into parasitic habits of dependence. Only through the nobler call of patriotism can our nation realise her highest ideals in thought and in action; to that call the nation will always respond. He had the inestimable privilege of winning the intimate friendship of Mr. G. K. Gokhale. Before leaving England, our foremost Indian statesman whose loss we so deeply mourn, had come to stay with the speaker for a few days at Eastbourne. He knew that this was to be their last meeting. Almost his parting question to Dr. Bose was whether science had anything to say about future incarnations. For himself, however he was certain that as soon as he would cast off his worn out frame he was to be born once more in the country he loved, and bear all the country that may be laid on him in her service. There can be no doubt that there must be salvation for a country which can count on sons as devoted as Gopal Krishna Gokhale.

—*Amrita Bazar Patrika,* 26-7-1915.

HISTORY OF A DISCOVERY

Substance of a Lecture delivered by Prof. J. C. Bose on the 20th November 1915, at the Ram Mohan Library, under the Presidency of the Hon'ble Mr. P. C. Lyon, and published at p. 693, Vol. xviii, of the "Modern Review" (July to December, 1915).

At the tournament held before the court at Hastinapur, more than twenty-five centuries ago, Karna, the reputed son of a Charioteer, had challenged the supremacy of Prince Arjuna. To this challenge Arjuna had returned a scornful answer; a prince could not cross swords with one who could claim no nobility of descent. "I am my own ancestor," replied Karna, and this perhaps the earliest assertion of the right of man to choose and determine his own destiny. In the realm of knowledge also the great achievements have been won only by men with determined purpose and without any adventitious aids. Undismayed by human limitations they had struggled in spite of many a failure. In their inquiry after truth they regarded nothing as too laborious, nothing too insignificant, nothing too painful. [Pg_155] This is the process which all must follow; there is no easier path.

The lecturer's research on the properties of Electric Waves was begun just twenty-one years ago. In this he was greatly encouraged by the appreciation shown by the Royal Society, which not only published his researches, but also offered a Parliamentary grant for the continuance of his work. The greatest difficulty lay in the construction of a receiver to detect invisible ether disturbances. For this a most laborious investigation had to be undertaken to find the action of electric radiation on all kinds of matter. As a result of this long and very patient work a new type of receiver was invented, so perfect in its action that the *Electrician* suggested its use in ships and electro-magnetic high houses for the communication and transmission of danger signals at sea through space. This was in 1895, several years in advance of the present wireless system. Practical application of the result of Dr. Bose's investigations appear so important that Great Britain and the United States granted him patents for his invention of a certain crystal receiver which proved to be the most sensitive detector of wireless signals.

UNIVERSAL SENSITIVENESS OF MATTER

In the course of his investigations Dr. Bose found [Pg_156] that the uncertainty of the early type of his receiver was brought on by fatigue, and that the curve of fatigue of his instrument closely resembled the fatigue curve of animal muscle. He was soon able to remove the 'tiredness' of his receiver by application of suitable stimulants; application of certain poisons, on the other hand, permanently abolished its sensitiveness. Dr. Bose was thus amazed at the discovery that inorganic matter was anything but inert, but that its particles were a thrill under the action of multitudinous forces that were playing on it. The lecturer was at this time constrained to choose whether to go on with the practical applications of his work, the success of which appeared to be assured, or to throw himself into a vortex of conflict for the establishment of some truth the glimmerings of which he was then but dimly beginning to perceive. It is very curious that the human mind is sometimes so constituted that it rejects lines of least resistance in favour of the more difficult path. Dr. Bose chose the more difficult path, and entered into a phase of activity which was to test all his strength.

CASTE IN SCIENCE

Dr. Bose's discovery of Universal sensitiveness of matter was communicated to the Royal Society on May 7th, 1901, when he himself gave a successful [Pg_157] experimental demonstration. His communication was, however, strongly assailed by Sir John Burden-Sanderson, the leading physiologist, and one or two of his followers. They had nothing to urge against his experiments but objected to a physicist straying into the preserve that had been specially reserved for the physiologist. He had unwittingly strayed into the domain of a new and unfamiliar caste system and offended its etiquette. In consequence of this opposition his paper, which was already in print, was not published. This is not by any means to be regarded as an injustice done to a stranger. Even Lord Rayleigh, who occupies an unique position in the world of science, was subjected to fierce attacks from the chemists, because he, a physicist, had ventured to predict that the air would be found to contain new elements not hitherto discovered.

It is natural that there should be prejudice against all innovations, and the attitude of Sir John Burden-Sanderson is easily explained. Unfortunately there was another incident about which similar explanation could not be urged. Dr. Bose's Paper had been placed in the archives of the Royal Society, so that technically there was no publication. And it came about that eight months after the reading of his Paper, another communication found publication in the Journal of a different society which [Pg_158] was practically the same as Dr. Bose's but without any acknowledgment. The author of this communication was a gentleman who had previously opposed him at the Royal Society. The plagiarism was subsequently discovered and led to much unpleasantness. It is not necessary to refer any more to the subject except as explanation of the fact that the determined hostility and misrepresentations of one man succeeded for more than ten years to bar all avenues of publication for his discoveries. But every cloud has its silver lining; this incident secured for him many true friends in England who stood for fair play, and whose friendship has proved to be a source of great encouragement to him.

FURTHER DIFFICULTIES

Dr. Bose's next work in 1903 was the discovery of the identity of response and of automatic activity in plant and animal and of the nervous impulse in plant. These new contributions were regarded as of such great importance that the Royal Society showed its special appreciation by recommending it to be published in their Philosophical transactions. But the same influence which had hitherto stood in his way triumphed once more, and it was at the very last moment that the publication was withheld. The Royal Society, however, informed him that his [Pg_159] results were of fundamental importance, but as they were so wholly unexpected and so opposed to the existing theories, that they would reserve their judgment until, at some future time, plants themselves could be made to record their answers to questions put to them. This was interpreted in certain quarters here as the final rejection of Dr. Bose's theories by the Royal Society, and the limited facilities which he had in the prosecution of his researches were in danger of being withdrawn. And everything was dark for him for the next ten years. The only thought that possessed him was how to make the plant give testimony by means of its own autograph.

LONG DELAYED SUCCESS

And when the night was at its darkest, light gradually appeared, and after innumerable difficulties had been overcome his Resonant Recorder was perfected, which enabled the plant to tell its own story. And in the meantime something still more wonderful came to pass. Hitherto all gates had been barred and he had to produce his passports everywhere. He now found friends who never asked him for credentials. His time had come at last. The Royal Society found his new methods most convincing and honoured him by publication of his researches in the Philosophical transactions. And his discoveries, [Pg_160] which had so long remained in obscurity, found enthusiastic acceptance.

Though his theories had thus received acceptance from the leading scientific men of the Royal Society, there was yet no general conviction of the identity of life reactions in plant and animal. No amount of controversy can remove the tendency of the human mind to follow precedents. The only thing left was to make the plant itself bear witness before the scientific bodies in the West, by means of self-records. At the recommendation of the Minister of Education, and of the Government of Bengal, the Secretary of State sanctioned his scientific deputation to Europe and America.

JOURNEY OF INDIAN PLANT ROUND THE WORLD

The special difficulty which he had to contend against lay in the fact that the only time during which the plant flourished at all in the West, was in the months of July and August, when the Universities and scientific societies were in vacation. The only thing left was to take the bold step of carrying growing plants from India and trust to human ingenuity to keep them alive during the journey. Four plants, two Mimosas and two Telegraph plants, were taken in a portable box with glass cover, and never let out of sight. In the [Pg_161] Mediterranean they encountered bitter cold for the first time and nearly succumbed. They were unhappier still in the Bay of Biscay, and when they reached London there was a sharp frost. They had to be kept in a drawing room lighted by gas, the deadly influence of which was discovered the next morning when all the plants were found to be apparently killed. Two had been killed, and the other two were brought round after much difficulty. The plants were at once transferred to the hot-house in Regents Park. For every demonstration in Dr. Bose's private Laboratory at Maida Vale, the plant had to be brought and returned in a taxicab with closed doors so that no sudden chill might kill them. When travelling, the large box in which they were, could not be trusted out of sight in the luggage van. They had practically to be carried in a reserved compartment. The unusual care taken of the box always roused the greatest curiosity, and in an incredibly short time large crowds would gather. When travelling long distances, for example from London to Vienna, the carriage accommodation had to be secured in advance. It was this that saved Dr. Bose from being interned in Germany, where he was to commence his lectures on the 4th August. He was to start for the University of Bonn on the 2nd, but on account of hasty mobilisation of troops in Germany he could not secure the reserved accommodation. [Pg_162] Two days after came the proclamation of War!

OUTCOME OF HIS WORK

The success of his scientific mission exceeded his most sanguine expectations. The work in which he long persevered in isolation and under most depressing difficulties, bore fruit at last. Apart from the full recognition that the progress of the world's science would be incomplete without India's special contributions, mutual appreciation and better understanding resulted from his visit. One of the greatest of Medical Institutions, the Royal Society of Medicine, has been pleased to regard his address before the society as one of the most important in their history and they expected that their science of medicine would be materially benefited by the researches that are being carried out by him in India. India has also been drawn closer to the great seats of learning in the West, to the Universities of Oxford and Cambridge; for there also the methods of inquiry initiated here have found the most cordial welcome. Many Indian students find their way to America, strangers in a strange land; hitherto they found few to advise and befriend them. It will perhaps be different now, since their leading Universities have begged from India the courtesy of hospitality for their post graduate scholars. Some [Pg_163] of these Universities again have asked for a supply of apparatus specially invented at Dr. Bose's laboratory which in their opinion will mark an epoch in scientific advance.

THE INEFFABLE WONDER BEHIND THE VEIL

As for the research itself, he said its bearings are not exclusively specialistic, but touch the foundation of various branches of science. To mention only a few; in medicine it had to deal with the fundamental reaction of protoplasm to various drugs, the solution of the problem why an identical agent brings about diametrically opposite effects in different constitutions; in the science of life it dealt with the new comparative physiology by which any specific characteristic of a tissue is traced from the simplest type in plant to the most complex in the animal; the study of the mysterious phenomenon of death and the accurate determination of the death point and the various conditions by which this point may be dislocated backwards and forwards; in psychology it had to deal with the unravelling of the great mystery that underlies memory and tracing it backwards to latent impressions even in the inorganic bodies which are capable of subsequent revival; and finally, the determination of the special characteristic of that vehicle through which sensiferous impulses are transmitted and the possibility [Pg_164] of changing the intensity and the tone of sensation. All these investigations, Dr. Bose said, are to be carried out by new physical methods of the utmost delicacy. He had in these years been able to remove the obstacles in the path and had lifted the veil so as to catch a glimpse of the ineffable wonder that had hitherto been hidden from view. The real work, he said, had only just begun.

A [Pg_165] SOCIAL GATHERING

At the Social Gathering held on the 16th December 1915, in the compound of the Calcutta Presidency College, to meet him after his highly successful tour through Europe, America and Japan, Dr. Bose spoke as follows: —

He said that it was his rare good fortune to have been amply rewarded for the hardships and struggles that he had gone through by the generous and friendly feelings of his colleagues and the love and trust of his pupils. He would say a few words regarding his experience in the Presidency College for more than three decades, which he hoped would serve to bring all who loved the Presidency College—present and past pupils and their teachers—in closer bonds of union. He would speak to them what he had learnt after years of patient labour, that the impossible became possible by persistent and determined efforts and adherence to duty and entire selflessness. The greatest obstacle often arises out of foolish misunderstanding of each other's ideals, such as the differing points of view, first of the Indian teacher, then of his western colleague, and [Pg_166] last but not least, the point of view of the Indian pupils themselves. In all these respects his experience had been wide and varied. He had both been an undergraduate and a graduate of the Calcutta University with vivid realization of an Indian student's aspirations; he had then become a student of conservative Cambridge and democratic London. And during his frequent visits to Europe and America he had become acquainted with the inner working of the chief universities of the world. Finally he had the unique privilege of being connected with the Presidency College for thirty-one years, from which no temptation could sever him. He had the deepest sense of the sacred vocation of the teacher. They may well be proud of a consecrated life—consecrated to what? To the guidance of young lives, to the making of men, to the shaping and determining of souls in the dawn of their existence, with their dreams yet to be realised.

Education in the West and in the East showed how different customs and ways might yet express a common ideal. In India the teacher was, like the head of a family, reverenced by his pupils so deeply as to show itself by touching the feet of their master. This in no servile act if we come to think of it; since it is the expression of the pupils' desire for his master's blessings, called down from heaven in an almost religious communion of souls. This consecration [Pg_167] is renewed every day, calling forth patient foresight of the teacher. As the father shows no special favour, but lets his love and compassion go out to the weakest, so it is with the Indian teacher and his pupil. There is the relation something very human, something very ennobling. He would say it was essentially human rather than distinctively Eastern. For do we not find something very like it in Mediaeval Europe? There too before the coming of the modern era with its lack of leisure and its adherence to system and machinery, there was a bond as sacred between the master and his pupils. Luther used to salute his class every morning with lifted hat, "I bow to you, great men of the future, famous administrators yet to be, men of learning, men of character who will take on themselves the burden of the world." Such is the prophetic vision given to the greatest of teachers. The modern teacher from England will set before him an ideal not less exalted—regarding his pupils as his comrades, he as an Englishman will instill into them greater virility and a greater public spirit. This will be his special contribution to the forming of our Indian youths.

Turning to the Indian students he could say that it was his good fortune never to have had the harmonious relation between teacher and pupils in any way ruffled during his long connection with them [Pg_168] for more than three decades. The real secret of success was in trying at times to see things from the student's point of view and to cultivate a sense of humour enabling him to enjoy the splendid self-assurance of youth with a feeling not unmixed with envy. In essential matters, however, one could not wish to meet a better type or one more quickly susceptible to finer appeals to right conduct and duty as Indian students. Their faults are rather of omission than of commission, since in his experience he formed that the moment they realised their teachers to be their friends, they responded instantly and did not flinch from any test, however severe, that could be laid on them.

— *The Presidency College Magazine. Vol. II, pages* 339-341.

LIGHT VISIBLE AND INVISIBLE

On the 14th January 1916, Dr. J. C. Bose delivered a public lecture, on Light Visible and Invisible, at the third Indian Science Congress held at Lucknow, before a crowded audience which included the Lieutenant-Governor (Sir James Meston).

Dr. Bose, in course of his lecture, spoke of the imperfection of our senses. Our ear, for example, fails to respond to all sounds. There are many sounds to which we are deaf. This was because our ear was tuned to answer to the narrow range of eleven octaves of sound vibrations. He showed a remarkable experiment of an artificial ear which remained irresponsive to various sounds, but when a particular note, to which it was tuned, was sounded even at the distant end of the hall, this ear picked it up and responded violently. As there were sounds audible and inaudible, so there were lights visible and invisible. The imperfection of our eye as a detector of ether vibrations was, however, far more serious. The eye could detect ether vibrations lying within a single octave—between 400 to 800 billion vibrations per second. Comparatively slow [Pg_170] vibrations of ether did not affect our eye and the disturbances they give rise to well-known as electric waves. The electric waves, predicted by Maxwell, were discovered by Hertz. These waves were about three metres long. They were about ten million times larger than the beams of visible light. Dr. Bose showed that the three short electric waves have the same property as a beam of light, exhibiting reflections, refraction, even total reflection, through a black crystal, double refraction, polarisation, and rotation of the plane of polarisation. The thinnest film of air was sufficient to produce total reflection of visible light with its extremely short wave lengths. But with the new electric waves which he produced, Dr. Bose showed that the critical thickness of air space determined by the refracting power of the prison and by the wave length of electric oscillations. Dr. Bose determined the index of refraction of electric waves for different materials, and eliminated a difficulty which presented itself in Maxwell's theory as to the relation between the index of refraction of light and the di-electric constant of insulators. He also measured the wave lengths of various oscillations. The order to produce short electric oscillations, to detect them and study their optical properties, he had to construct a large number of instruments. It was a hard task to produce very short electric waves which had [Pg_171] enough energy to be detected, but Dr. Bose overcame this difficulty by constructing radiators or oscillators of his own type, which emitted the shortest waves with sufficient energy. As a receiver he used a sensitive metallic coherer, which in itself led to new and important discoveries. When electric waves fall on a loose contact between two pieces of metals, the resistance of the contact changes and a current passes through the contact indicating the existence of electrical oscillations. Dr. Bose discovered the surprising fact that with potassium metal the resistance of the contact increases under the action of electric waves and that this contact exhibits an automatic recovery. He found further that the change of the metallic contact resistance when acted upon by electric waves, is a function of the atomic weight. These phenomena led to a new theory of metallic coherers. Before these discoveries it was assumed that the particles of the two metallic pieces in contact are, as it were, fused together, so that the resistance decreases. But the increasing resistance appearing for some elements, led to the theory that the electric forces in the waves produced a peculiar molecular action or a re-arrangement of the molecules, which may either increase or decrease the contact resistance.

—*Pioneer*,—16-1-1916.

HINDU UNIVERSITY ADDRESS

The foundation of the Hindu University was laid by Lord Hardinge on the 4th February 1916. "Many striking addresses were delivered on the occasion. Professor J.C. Bose in his masterly address went to the root of the matter and pointed in an inspiring manner what should be done to make the Hindu University worthy of its name. He deprecated a repetition of the Universities of the West." He said:—

In tracing the characteristic phenomenon of life from simple beginnings in that vast region which may be called unvoiced, as exemplified in the world of plants, to its highest expression in the animal kingdom, one is repeatedly struck by the one dominant fact that in order to maintain an organism at the height of its efficiency something more than a mechanical perfection of its structure is necessary. Every living organism, in order to maintain its life and growth, must be in free communion with all the forces of the Universe about it.

STIMULUS [Pg_173] WITHIN AND WITHOUT

Further, it must not only constantly receive stimulus from without, but must also give out something from within, and the healthy life of the organism will depend on these two fold activities of inflow and outflow. When there is any interference with these activities, then morbid symptoms appear, which ultimately must end in disaster and death. This is equally true of the intellectual life of a Nation. When through narrow conceit a Nation regards itself self-sufficient and cuts itself from the stimulus of the outside world, then intellectual decay must inevitably follow.

SPECIAL FUNCTION OF A NATION

So far as regards the receptive function. Then there is another function in the intellectual life of a Nation, that of spontaneous outflow, that giving out of its life by which the world is enriched. When the Nation has lost this power, when it merely receives, but cannot give out, then its healthy life is over, and it sinks into a degenerate existence which is purely parasitic.

HOW INDIA CAN TEACH

How can our Nation give out of the fulness of the life that is in it, and how can a new Indian [Pg_174] University help in the realisation of this object? It is clear that its power of directing and inspiring will depend on its world status. This can be secured to it by no artificial means, nor by any strength in the past; and what is the weakness that has been paralysing her activities for the accomplishment of any great scientific work? There must be two different elements, and these must be evenly balanced. Any excess of either will injure it.

HOW TO SECURE THIS STATUS

This world status can only be won by the intrinsic value of the great contributions to be made by its own Indian scholars for the advancement of the world's knowledge. To be organic and vital our new University must stand primarily for self-expression, and for winning for India a place she has lost. Knowledge is never the exclusive possession of any particular race, nor does it recognise geographical limitations. The whole world is interdependent, and a constant stream of thought had been carried out throughout the ages enriching the common heritage of mankind. Although science was neither of the East nor of the West but international, certain aspects of it gained richness by reason of their place of origin.

In any case if India need to make any contribution [Pg_175] to the world it should be as great as the hope they cherished for her. Let them not talk of the glories of the past till they have secured for her, her true place among the intellectual nations of the world. Let them find out how she had fallen from her high estate and ruthlessly put an end to all that self satisfied and little-minded vanity which had been the cause of their fatal weakness. What was it that stood in her way? Was her mind paralysed by weak superstitious fears? That was not so; for her great thinkers, the Rishis, always stood for freedom of intellect and while Galileo was imprisoned and Bruno burnt for their opinions, they boldly declared that even the Vedas were to be rejected if they did not conform to truth. They urged in favour of persistent efforts for the discovery of physical causes yet unknown, since to them nothing was extra-physical but merely mysterious because of a hitherto unascertained cause. Were they afraid that the march of knowledge was dangerous to true faith? Not so. For their knowledge and religion were one.

These are the hopes that animate us. For there is something in the Hindu culture which is possessed of extraordinary latent strength by which it has resisted the ravages of time and the destructive changes which have swept over the earth. And indeed a capacity to endure through infinite transformations [Pg_176] must be innate in that mighty civilisation which has seen the intellectual culture of the Nile Valley, of Assyria and of Babylon war and wane and disappear and which to-day gazes on the future with the same invincible faith with which it met the past.

—*Modern Review, vol. XIX, pages 277, 278.*

THE HISTORY OF A FAILURE THAT WAS GREAT

At the invitation of the President and the committee of the Faridpore Industrial Exhibition, Dr. J. C. Bose gave a lecture on the life of his father, the late Babu Bhugwan Chunder Bose, who founded the Exhibition at Faridpore, where he was the sub-divisional officer, 50 years ago. It was published in the Modern Review for February 1917—volume xxi, p. 221. In course of his address, said Dr. Bose:—

It is the obvious, the insistent, the blatant that often blinds us to the essential. And in solving the mystery that underlies life, the enlightenment will come not by the study of the complex man, but through the simpler plant. It is the unsuspected forces, hidden to the eyes of men,—the forces imprisoned in the soil and the stimuli of alternating flash of light and the gloomings of darkness these and many others will be found to maintain the ceaseless activity which we know as the fulness of throbbing life.

This is likewise true of the congeries of life which we call a society or a nation. The energy which [Pg_178] moves this great mass in ceaseless effort to realise some common aspiration, often has its origin in the unknown solitudes of a village life. And thus the history of some efforts, not forgotten, which emanated from Faridpore, may be found not unconnected with which India is now meeting her problems to-day. How did these problems first dawn in the minds of some men who forecast themselves by half a century? How fared their hopes, how did their dreams become buried in oblivion? Where lies the secret of that potency which makes certain efforts apparently doomed to failure, rise renewed from beneath the smouldering ashes? Are these dead failures, so utterly unrelated to some great success that we may acclaim to day? When we look deeper we shall find that this is not so, that as inevitable as in the sequence of cause and effect, so unrelenting must be the sequence of failure and success. We shall find that the failure must be the antecedent power to lie dormant for the long subsequent dynamic expression in what we call success. It is then and then only that we shall begin to question ourselves which is the greater of the two, a noble failure or a vulgar success.

As a concrete example, I shall relate the history of a noble failure which had its setting in this little corner of the earth. And if some of the audience thought that the speaker has been blessed with life [Pg_179] that has been unusually fruitful, they will soon realise that the power and strength that nerved me to meet the shocks of life were in reality derived at this very place, where I witnessed the struggle which overpowered a far greater life.

STIMULUS OF CONTACT WITH WESTERN CULTURE

An impulse from outside reacts on impressionable bodies in two different ways, depending on whether the recipient is inert or fully alive. The inert is fashioned after the pattern of the impression made on it, and this in infinite repetition of one mechanical stamp. But when an organism is fully alive, the answering reaction is often of an altogether different character to the impinging stimulus. The outside shocks stir up the organism to answer feebly or to utmost in ways as multitudinous and varied as life itself. So the first impetus of Western education impressed itself on some in a dead monotony of imitation of things Western; while in others it awakened all that was greatest in the national memory. It is the release of some giant force which lay for long time dormant. My father was one of the earliest to receive the impetus characteristic of the modern epoch as derived from the West. And in his case it came to pass that the stimulus evoked the latent potentialities of his race for evolving [Pg_180] modes of expression demanded by the period of transition in which he was placed. They found expression in great constructive work, in the restoration of quiet amidst disorder, in the earliest effort to spread education both among men and women, in questions of social welfare, in industrial efforts, in the establishment of people's Bank and in the foundation of industrial and technical schools. And behind all these efforts lay a burning love for his country and its nobler traditions.

MATTERS EDUCATIONAL

In educational matters he had very definite ideas which is now becoming more fully appreciated. English schools were at that time not only regarded as the only efficient medium for instruction. While my father's subordinates sent their children to the English schools intended for gentle folks, I was sent to the vernacular school where my comrades were hardy sons of toilers and of others who, it is now the fashion to regard, were belonging to the depressed classes. From these who tilled the ground and made the land blossom with green verdure and ripening corn, and the sons of the fisher folk, who told stories of the strange creatures that frequented the unknown depths of mighty rivers and stagnant pools, I first derived the lesson of that which constitutes true manhood. From them too I drew my love of [Pg_181] nature. When I came home accompanied by my comrades I found my mother waiting for us. She was an orthodox Hindu, yet the "untouchableness" of some of my school fellows did not produce any misgivings in her. She welcomed and fed all these as her own children; for it is only true of the mother heart to go out and enfold in her protecting care all those who needed succour and a mother's affection. I now realise the object of my being sent at the most plastic period of my life to the vernacular school, where I was to learn my own language, to think my own thoughts and to receive the heritage of our national culture through the medium of our own literature. I was thus to consider myself one with the people and never to place myself in an equivocal position of assumed superiority. This I realised more particularly when later I wished to go to Europe and to compete for the Indian Civil Service, his refusal as regards that particular career was absolute. I was to rule nobody but myself, I was to be a scholar not an administrator.

THE HISTORY OF A FAILURE THAT WAS GREAT

There has been some complaint that the experiment of meeting out cut and dried moral texts as a part of school routine has not proved to be so effective as was expected by their promulgators. [Pg_182] The moral education which we received in our childhood was very indirect and came from listening to stories recited by the

'Kathas' on various incidents connected with our great epics. Their effect on our minds was very great; this may be because our racial memory makes us more prone to respond to certain ideals that have been impressed on the consciousness of the nation. These early appeals to our emotions have remained persistent; the only difference is that which was there as a narrative of incidents more or less historical, is now realised as eternally true, being an allegory of the unending struggle of the human soul in its choice between what is material and that other something which transcends it. The only pictures now in my study are a few frescoes done for me by Abanindra Nath Tagore and Nanda Lal Bose. The first fresco represents Her, who is the Sustainer of the Universe. She stands pedestalled on the lotus of our heart. The world was at peace; but a change has come. And She under whose Veil of Compassion we had been protected so long, suddenly flings us to the world of conflict. Our great epic, the Mahabharata, deals with this great conflict, and the few frescoes delineate some of the fundamental incidents. The coming of the discord is signalled by the rattle of dice, thrown by Yudhisthira, the pawn at stake, being the crown. Two hostile arrays are set in [Pg_183] motion, mighty Kaurava armaments meeting in shock of battle the Pandava host with Arjuna as the leader, and Krishna as his Divine Charioteer. At the supreme moment Arjuna had flung down his earthly weapon, Gandiva. It was then that the eternal conflict between matter and spirit was decided. The next panel shows the outward or the material aspect of victory. Behind a foreground of waving flags is seen the battle field of Kurukshetra with procession of white-clad mourning women seen by fitful lights of funeral pyres. In the last panel is seen Yudhisthira renouncing the fruits of his victory setting out on his last journey. In front of him lies the vast and sombre plain and mountain peaks, faintly visible by gleams of unearthly light, unlocalised but playing here and there. His wife and his brothers had fallen behind and dropped one by one. There is to be no human companion in his last journey. The only thing that stood by him and from which he had never been really separated is Dharma or the Spirit of Righteousness.

LIFE OF ACTION

Faridpur at that time enjoyed a notoriety of being the stronghold of desperate characters, dacoits by land and water. My father had captured single-handed one of the principal leaders, whom he sentenced to a long term of imprisonment. After [Pg_184] release he came to my father and demanded some occupation, since the particular vocation in which he had specialised was now rendered impossible. My father took the unusual course to employ him as my special attendant to carry me, a child of four, on his back to the distant village school. No nurse could be tenderer than this ex-leader of lawless men, whose profession had been to deal out wounds and deaths. He had accepted a life of peace but he could not altogether wipe out his old memories. He used to fill my infant mind with the stories of his bold adventures, the numerous fights in which he had taken part, the death of his companions and his hair-breadth escapes. Numerous were the decorations he bore. The most conspicuous was an ugly mark on his breast left by an arrow and a hole on the thigh caused by a spear thrust. The trust imposed on this marauder proved to be not altogether ill placed for once in a river journey we were pursued by several long boats filled with armed dacoits. When these boats came too near for us to effect an escape the erstwhile dacoit leader, my attendant, stood up and gave a peculiar cry, which was evidently understood. For the pursuing boats vanished at the signal.

INDUSTRIAL EFFORTS

I come now to another period of his life fifty years from now, when he foresaw the economic danger [Pg_185] that threatened his country. This Agricultural and Industrial Exhibition was one of the first means he thought of to avert the threatened danger. Here also he attempted to bring together other activities. Evening entertainments were given by the performances of "Jatras," which have been the expression of our national drama and which have constantly enriched our Bengali literature by the contributions of village bards and composers. There were athletic tournaments also and display of physical strength and endurance. He also established here the people's Bank, which is now in a most flourishing condition. He established industrial and technical schools, and it was there that the inventive bend of my mind received its first impetus. I remember the deep impression made on my mind by the form of worship rendered by the artisans to Viswakarma God in his aspect as the Great Artificer: His hand it was that was moulding the whole creation; and it seemed that we were the instruments in his hand, through whom he intended to fashion some Great Design.

In practical agriculture my father was among Indians one of the first to start a tea industry in Assam, now regarded as one of the most flourishing. He gave practically everything in the starting of some Weaving Mills. He stood by this and many other efforts in industrial developments. The success of [Pg_186] which I spoke did not come till long after—too late for him to see it. He had come before the country was ready, and it happened to him as it must happen to all pioneers. Every one of his efforts failed and the crash came. And a great burden fell on us which was only lifted by our united effects just before his work here was over.

A failure? Yes but not ignoble or altogether futile. Since it was through the witnessing of this struggle that the son learned to look on success or failure as one, to realise that some defeat was greater than victory. And if my life in any way proved to be fruitful, then that came through the realisation of this lesson.

To me his life had been one of blessing and daily thanksgiving. Nevertheless every one had said that he wrecked his life which was meant for far greater things. Few realise that out of the skeletons of myriad lives have been built vast continents. And it is on the wreck of a life like his and of many such lives there will be built the Greater India yet to be. We do not know why it should be so, but we do know that the Earth Mother is hungry for sacrifice.

QUEST OF TRUTH AND DUTY

Sir Jagadis Chandra Bose delivered the following Address, on the 25th February 1917, to the students of the Presidency College on receiving their *Arghya* and congratulations on the occasion of his knighthood. It was published in the Modern Review for March 1917 — Volume XXI, p. 343.

In your congratulations for the recent honour, you have overlooked a still greater that came to me a year ago, when I was gazetted as your perpetual professor, so that the tie which binds me to you is never to be severed. Thirty-two years ago I sought to be your teacher. For the trust that you imposed on me could I do anything less than place before you the highest that I knew? I never appealed to your weaknesses but your strength. I never set before you that was easy but used all the compulsion for the choice of the most difficult. And perhaps as a reward for these years of effort I find all over India those who have been my pupils occupying positions of the highest trust and responsibility in different walks of life. I do not merely count those who have won fame and success but I also claim [Pg_188] many others who have taken up the burden of life manfully and whose life of purity and unselfishness has brought gleams of joy in suffering lives.

THE LAW UNIVERSAL

Through science I was able to teach you how the seeming veils the real; how though the garish lights dazzle and blind us, there are lights invisible, which glow persistently after the brief flare burns out. One came to realise how all matter was one, how unified all life was. In the various expressions of life even in the realm of thought the same Universal law prevails. There was no such thing as brute matter, but that spirit suffused matter in which it was enshrined. One also realised dimly a mysterious Cyclic Law of Change, seen not merely in inorganic matter but also in organised life and its highest manifestations. One saw how inertness passes into the climax of activity and how that climax is perilously near its antithetic decline. This basic change puzzles us by its seeming caprice not merely in our physical instruments but also in the cycle of individual life and death and in the great cycle of the life and death of nations. We fail to see things in their totality and we erect barriers that keep kindreds apart. Even science which attempts to rise above common limitations, has not escaped the doom which limited vision imposes. We have [Pg_189] caste in science as in religion and in politics, which divides one into conflicting many. The law of Cyclic change follows us relentlessly even in the realm of thought. When we have raised ourselves to the highest pinnacle, through some oversight we fall over the precipice. Men have offered their lives for the establishment of truth. A climax is reached after which the custodians of knowledge themselves bar further advance. Men who have fought for liberty impose on themselves and on others the bond of slavery. Through centuries have men striven to erect a mighty edifice in which Humanity might be enshrined; through want of vigilance the structure crumbled into dust. Many cycles must yet be run and defeats must yet be borne before man will establish a destiny which is above change.

And through science I was able to teach you to seek for truth and help to discover it yourself. This attitude of detachment may possess some advantage in the proper understanding of your duties. You will have, besides, the heritage of great ideals that have been handed down to you. The question which you have to decide is duty to yourself, to the king and to your country. I shall speak to you of the ideals which we cherish about these duties.

DUTY TO SELF

As regards duty to self, can there be anything so [Pg_190] inclusive as being true to your manhood? Stand upright and do not be either cringing or vulgarly self-assertive. Be righteous. Let your words and deeds correspond. Lead no double life. Proclaim what you think right.

IDEAL OF KINGSHIP

The Indian ideal of kingship will be clear to you if I recite the invocation with which we crowned our kings from the Vedic Times:

"Be with us. We have chosen thee
Let all the people wish for thee
Stand steadfast and immovable
Be like a mountain unremoved
And hold thy kingship in thy grasp."

We have chosen thee, our prayers have consecrated thee, for all the wishes of the people went with thee. Thou art to stand as mountain unremoved, for thy throne is planted secure on the hearts of thy people. Stand steadfast then, for we have endowed thee with power irresistible. Fall therefore not away; but let thy sceptre be held firmly in thy grasp.

Which is more potent, Matter or Spirit? Is the power with which the people endow their king identical with the power of wealth with which we enrich him by paying him his Royal dues? We make him irresistible not by wealth but by [Pg_191] the strength of our lives, the strength of our mind, may, we have to pay him more according to our ancient Lawgivers, in as much as the eighth part of our deeds and virtues, and the merit we have ourselves acquired. We can only make him irresistible by the strength of our lives, the strength of our minds, and the strength that comes out of righteousness.

DUTY TO OUR COUNTRY

And lastly, what are our duties to our country? These are essentially to win honour for it and also win for it security and peace. As regards winning honour for our country, it is true that while India has offered from the earliest times welcome and hospitality to all peoples and nationalities her children have been subjected to intolerable humiliations in other countries even under the flag of our king.

There can be no question of the fundamental duty of every Indian to stand up and uphold the honour of his country and strove for the removal of wrong.

The general task of redressing wrong is not a problem of India alone, but one in which the righteous men are interested the world over. For wrong cries for redress everywhere, in the clashings interests of the rich and poor, between capital and labour, between those who hold the power and those from whom it has been withheld, — in a word in the struggle of the Disinherited.

[Pg_192] When any man is rendered unable to uphold his manhood and self-respect and woman are deprived of the chivalrous protection and consideration of men and subjected to degradation, the general level of manhood or womanhood in the world is lowered. It then becomes an outrage to humanity and a challenge to all men to safeguard the sacredness of our common human nature.

What is the machinery which sets a going a world movement for the redress of wrong? For this I need not cite instances from the history of other countries but take one which is known to you and in which the living actors are still among us. In the midst of the degradation of his countrymen in South Africa, there stood up a man himself nurtured in luxury, to take up the burden of the disinherited. His wife too stood by him, a lady of gentle birth. We all know who that man is — he is Gandhi, — and what humiliations and suffering he went through. Do you think he suffered in vain and that his voice remained unheard? It was not so, for in the great vortex of passion for Justice, there were caught others — men like Polak and Andrews. Are they your countrymen? Not in the narrow sense of the word but truly in a larger sense, that these who choose to bear and suffer belong to one clan the clan from which Kshatriya Chivalry is recruited. The removal of suffering and of the cause of suffering is [Pg_193] the Dharma of the strong Kshatriya. The earth is the wide and universal theatre of man's woeful pageant. The question is who is to suffer more than his share. Is the burden to fall on the weak or the strong? Is it to be under hopeless compulsion or of voluntary acceptance?

DEFENCE OF HOMELAND

In your services for your country there is no higher at the present moment than to ensure for her security and peace. We have so long enjoyed the security of peace without being called upon to maintain it. But this is no longer so.

At no time within the recent history of India has there been so quick a readjustment and appreciation as regards proper understanding of the aspiration of the Indian people. This has been due to what India has been able to offer not merely in the regions of thought but also in the fields of battle.

MASS RESPONSE

And remember that when the world is in conflagration, this corner which has hitherto escaped it, will not evade the peril which threatens it. The march of disaster will then be terribly rapid. You have soon to prepare yourself against any hostile sides. You can only withstand it if the whole people realise the imminent danger. You can by your [Pg_194] thought and by your action awaken and influence the multitude. Do not have any misgivings about the want of long previous preparations. Have you not already seen how mind triumphs over matter and have not some of you with only a few months' preparation stood fearless at your post in Mesopotamia and won recognition by your calm collectedness and true heroism? They may say that you are but a small handful, what of the vast illiterate millions? Illiterate in what sense? Have not the ballads of these illiterates rendered into English by our Poet touched profoundly the hearts of the very elect of the West? Have not the stories of their common life appealed to the common kinship of humanity? If you still have some doubts about the power of the multitude to respond instantly to the call of duty, I shall relate an incident which came within my own personal experience. I had gone on a scientific expedition to the borders of the Himalayan terrai of Kumaun; a narrow ravine was between me and the plateau on the other side. Terror prevailed among the villagers on the other side of the ravine; for a tigress had come down from the forest. And numerous had been the toll in human lives exacted. Petitions had been sent up to the Government and questions had been asked in Parliament. A reward of Rs. 500 had been offered. Various captains in the army with battery of guns [Pg_195] came many a time, but the reward remained unclaimed. The murderess of the forest would come out even in broad day-light and leisurely take her victims from away their companions. Nothing could circumvent her demoniac cunning. When all hopes had nearly vanished, the villagers went to Kaloo Singh, who possessed an old matchlock. At the special sanction of the Magistrate he was allowed to buy a quantity of gunpowder; the bullets he himself made by melting bits of lead. With his primitive weapon with the entreaties of his villagers ringing in his ears Kaloo Singh started on his perilous journey. At midday I was startled by the groanings of some animals in pain. The tigress had sprung among a herd of buffalo and with successive strokes of its mighty paws had killed two buffaloes and left them in the field. Kaloo Singh waited there for the return of the tigress to the kill. There was not a tree near by; only there was a low bush behind which he lay crouched. After hours of waiting as the sun was going down he was taken aback by the sudden apparition of the tigress which stood within six feet of him. His limbs had become half paralysed from cold and his crouching position. Trying to raise his gun he could take no aim as his arm was shaking with involuntary fear. Kaloo Singh explained to me afterwards how he succeeded in shaking off his mortal terror. "I quietly said to myself, [Pg_196] Kaloo Singh, Kaloo Singh, who sent you here? Did not the villagers put their trust on you! I could then no longer lie in hiding, and I stood up and something strange and invigorating crept up strength into my body. All the trembling went and I became as hard as steel. The tigress had seen me and with eyes blazing crouched for the spring lashing its tail. Only six feet lay between. She sprang and my gun also went off at the same time and she missed her aim and fell dead close to me." That was how a common villager went off to meet death at the call of something for which he could give no name and the mother and wife of Kaloo Singh had also bidden him go. There are millions of Kaloo Singhs with mother and sisters and wife to send them forth. And you too have many loved ones who would themselves bid you arm for the defence of your homes.

DIFFERENCE OF TEMPERAMENT

The issue is clear, and immediate action is imperative. But action is delayed by misunderstanding arising out of temperamental differences between the Governing Class and the People. Curiously enough the respective responsive characteristics of the Anglo Saxon and the Indians are paralleled by the two types of responses seen in all living matter. In the one type the response is slow but proportionate to the stimulus that excites it. The response [Pg_197] grows with the strength of external force. In the other it is quite different—here it is an all-or-none principle. It either responds to the utmost or nothing at all. This is also illustrated in the different racial characteristics. The Anglo Saxon has even by his rights by struggle, step by step. The insignificant little has, by accumulation, became large, and which has been gained, has been gained for all time. But in the Indian the ideal and the emotional are the only effective stimulus. The ideal of his King is Rama, who renounced his kingdom and even his beloved for an idea. One day a king and another day a bare-footed wanderer in the forest! Who cares? All or nothing!

The concessions made by a modern form of Government safeguarded by necessary limitations may appear almost as grudging gifts. The Indian wants something which comes with unhesitating frankness and warmth and strikes his ideality and imagination. But ancient and modern kingship are sometimes at one in direct and spontaneous pronouncement of the royal sympathy. Such was the Proclamation of Queen Victoria which stirred to its depths the popular heart.

"In the Prosperity of Our subjects will be our strength, in their contentment Our security, in their Gratitude Our best Reward."

That there are increasingly frequent reflexes in [Pg_198] our Government to popular needs and wishes is happily illustrated at a most opportune moment from the statements in the recent *Gazette of India* and cables received from London. In the former we find that the Viceroy and his council had recommended the abolition of the system of indentured labour. In the telegram from London Mr. Chamberlain states that the Viceroy has informed him that Indians will be eligible for commissions in the New Defence of India Army.

MARCH OF WORLD TRAGEDY

In the meantime the Embodiment of World Tragedy is marching with giant strides. Brief will be his hesitation whether he will choose to step first to the East or to the West. Already across the Atlantic, they are preparing for the dreaded visitation. In the farthest East they have long been prepared. We alone are not ready. Pity for our helplessness will not stay the impending disaster, rather provoke it. When that comes, as assuredly it will unless we are prepared to resist, havoc will be let loose and horrors perpetrated before which the imagination quails back in dismay.

I have tried to lay before you as dispassionately as I could the issues involved. But some of you may cry out and say, we can not live in cold scientific and philosophic abstractions. Emotion is more [Pg_199] to us than pure reasoning. We cannot stay in this indecision which is paralysing our wills and crushing the soul out of us. The world is offering their best and behold them marching to be immolated so that by the supreme offering of death they might win safety and honor for their motherland. There is no time for wavering. We too will throw in our lot with those who are fighting. They say that by our lives we shall win for our birth-land an honoured place in their federation. We shall trust them. We shall stand by their side and fight for our home and homeland. And let Providence shape the Issue.

THE VOICE OF LIFE

The following is the Inaugural Address delivered by Sir J. C. Bose, on the 30th November 1917, in dedicating the Bose Institute to the Nation.

I dedicate to-day this Institute—not merely a Laboratory but a Temple. The power of physical methods applies for the establishment of that truth which can be realised directly through our senses, or through the vast expansion of the perceptive range by means of artificially created organs. We still gather the tremulous message when the note of the audible reaches the unheard. When human sight fails, we continue to explore the region of the invisible. The little that we can see is as nothing compared to the vastness of that which we cannot. Out of the very imperfection of his senses man has built himself a raft of thought by which he makes daring adventures on the great seas of the Unknown. But there are other truths which will remain beyond even the supersensitive methods known to science. For these we require faith, tested not in a few years but by an entire life. And a temple is erected as a fit memorial for the establishment of that truth for which faith was needed. The [Pg_201] personal, yet general, truth and faith whose establishment this Institute commemorates is this: that when one dedicates himself wholly for a great object, the closed doors shall open, and the seemingly impossible will become possible for him.

Thirty-two years ago I chose teaching of science as my vocation. It was held that by its very peculiar constitution, the Indian mind would always turn away from the study of Nature to metaphysical speculations. Even had the capacity for inquiry and accurate observation been assumed present, there were no opportunities for their employment; there were no well-equipped laboratories nor skilled mechanicians. This was all too true. It is for man not to quarrel with circumstances but bravely accept them; and we belong to that race and dynasty who had accomplished great things with simple means.

FAILURE AND SUCCESS

This day twenty-three years ago, I resolved that as far as the whole-hearted devotion and faith of one man counted, that would not be wanting and within six months it came about that some of the most difficult problems connected with Electric Waves found their solution in my Laboratory and received high appreciation from Lord Kelvin, Lord Rayleigh and other leading physicists. The Royal Society [Pg_202] honoured me by publishing my discoveries and offering, of their own accord, an appropriation from the special Parliamentary Grant for the advancement of knowledge. That day the closed gates suddenly opened and I hoped that the torch that was then lighted would continue to burn brighter, and brighter. But man's faith and hope require repeated testing. For five years after this, the progress was interrupted; yet when the most generous and wide appreciation of my work had reached almost the highest point there came a sudden and unexpected change.

LIVING AND NON-LIVING

In the pursuit of my investigations I was unconsciously led into the border region of physics and physiology and was amazed to find boundary lines vanishing and points of contact emerge between the realms of the Living and Non-living. Inorganic matter was found anything but inert; it also was a thrill under the action of multitudinous forces that played on it. A universal reaction seemed to bring together metal, plant and animal under a common law. They all exhibited essentially the same phenomena of fatigue and depression, together with possibilities of recovery and of exaltation, yet also that of permanent irresponsiveness which is associated with death. I was filled with awe at this stupendous [Pg_203] generalisation; and it was with great hope that I announced my results before the Royal Society,—results demonstrated by experiments. But the physiologists present advised me, after my address, to confine myself to physical investigations in which my success had been assured, rather than encroach on their preserve. I had thus unwittingly strayed into the domain of a new and unfamiliar caste system and so offended its etiquette. An unconscious theological bias was also present which confounds ignorance with faith. It is forgotten that He, who surrounded us with this ever-evolving mystery of creation, the ineffable wonder that lies hidden in the microcosm of the dust particle, enclosing within the intricacies of its atomic form all the mystery of the cosmos, has also implanted in us the desire to question and understand. To the theological bias was added the misgivings about the inherent bent of the Indian mind towards mysticism and unchecked imagination. But in India this burning imagination which can extort new order out of a mass of apparently contradictory facts, is also held in check by the habit of meditation. It is this restraint which confers the power to hold the mind in pursuit of truth, in infinite patience, to wait, and reconsider, to experimentally test and repeatedly verify.

It is but natural that there should be prejudice, even in science, against all innovations; and I was [Pg_204] prepared to wait till the first incredulity could be overcome by further cumulative evidence. Unfortunately there were other incidents and misrepresentations which it was impossible to remove from this insulating distance. Thus no conditions could have been more desperately hopeless than those which confronted me for the next twelve years. It is necessary to make this brief reference to this period of my life; for one who would devote himself to the search of truth must realise that for him there awaits no easy life, but one of unending struggle. It is for him to cast his life as an offering, regarding gain and loss, success and failure, as one. Yet in my case this long persisting gloom was suddenly lifted. My scientific deputation in 1914, from the Government of India, gave the opportunity of giving demonstrations of my discoveries before the leading scientific societies of the world. This led to the acceptance of my theories and results, and the recognition of the importance of the Indian contribution to the advancement of the world's science. My own experience told me how heavy, sometimes even crushing, are the difficulties which confront an inquirer here in India; yet it made me stronger in my determination, that I shall make the path of those who are to follow me less arduous, and that India, is never to relinquish what has been won for her after years of struggle.

THE TWO IDEALS

What is it that India is to win and maintain? Can anything small or circumscribed ever satisfy the mind of India? Has her own history and the teaching of the past prepared her for some temporary and quite subordinate gain? There are at this moment two complementary and not antagonistic ideals before the country. India is drawn into the vortex of international competition. She has to become efficient in every way, — through spread of education, through performance of civic duties and responsibilities, through activities both industrial and commercial. Neglect of these essentials of national duty will imperil her very existence; and sufficient stimulus for these will be found in success and satisfaction of personal ambition.

But these alone do not ensure the life of a nation. Such material activities have brought in the West their fruit, in accession of power and wealth. There has been a feverish rush even in the realm of science, for exploiting applications of knowledge, not so often for saving as for destruction. In the absence of some power of restraint, civilisation is trembling in an unstable poise on the brink of ruin. Some complementary ideal there must be to save man from that mad rush which must end in disaster. He has followed the lure and excitement of some insatiable ambition, never pausing for a moment to think of [Pg_206] the ultimate object for which success was to serve as a temporary incentive. He forgot that far more potent than competition was mutual help and co-operation in the scheme of life. And in this country through milleniums, there always have been some who, beyond the immediate and absorbing prize of the hour, sought for the realisation of the highest ideal of life — not through passive renunciation, but through active struggle. The weakling who has refused the conflict, having acquired nothing has nothing to renounce. He alone who has striven and won, can enrich the world by giving away the fruits of his victorious experience. In India such examples of constant realisation of ideals through work have resulted in the formation of a continuous living tradition. And by her latent power of rejuvenescence she has readjusted herself through infinite transformations. Thus while the soul of Babylon and the Nile Valley have transmigrated, ours still remains vital and with capacity of absorbing what time has brought, and making it one with itself.

The ideal of giving, of enriching, in fine, of self-renunciation in response to the highest call of humanity is the other and complementary ideal. The motive power for this is not to be found in personal ambition but in the effacement of all littlenesses, and uprooting of that ignorance which regards anything as gain which is to be purchased [Pg_207] at others' loss. This I know, that no vision of truth can come except in the absence of all sources of distraction, and when the mind has reached the point of rest.

Public life, and the various professions will be the appropriate spheres of activity for many aspiring young men. But for my disciples, I call on those very few, who, realising inner call, will devote their whole life with strengthened character and determined purpose to take part in that infinite struggle to win knowledge for its own sake and see truth face to face.

ADVANCEMENT AND DIFFUSION OF KNOWLEDGE

The work already carried out in my laboratory on the response of matter, and the unexpected revelations in plant life, foreshadowing the wonders of the highest animal life, have opened out very extended regions of inquiry in Physics, in physiology in Medicine, in Agriculture and even in Psychology. Problems, hitherto regarded as insoluble, have now been brought within the sphere of experimental investigation. These inquiries are obviously more extensive than those customary either among physicists or physiologists, since demanding interests and aptitudes hitherto more or less divided between them. In the study of Nature, there is a [Pg_208] necessity of the dual view point, this alternating yet rhythmically unified interaction of biological thought with physical studies, and physical thought with biological studies. The future worker with his freshened grasp of physics, his fuller conception of the inorganic world, as indeed thrilling with "the promise and potency of life" will redouble his former energies of work and thought. Thus he will be in a position to win now the old knowledge with finer sieves, to research it with new enthusiasm and subtler instruments. And thus with thought and toil and time he may hope to bring fresher views into the old problems. His handling of these will be at once more vital and more kinetic, more comprehensive and unified.

The farther and fuller investigation of the many and ever-opening problems of the nascent science which includes both Life and Non-Life are among the main purposes of the Institute I am opening to-day; in these fields I am already fortunate in having a devoted band of disciples, whom I have been training for the last ten years. Their number is very limited, but means may perhaps be forthcoming in the future to increase them. An enlarging field of young ability may thus be available, from which will emerge, with time and labour, individual originality of research, productive invention and some day even creative genius.

[Pg_209] But high success is not to be obtained without corresponding experimental exactitude, and this is needed to-day more than ever, and to-morrow yet more again. Hence the long battery of supersensitive instruments and apparatus, designed here, which stand before in their cases in our entrance hall. They will tell you of the protracted struggle to get behind the deceptive seeming into the reality that remained unseen;—of the continuous toil and persistence and of ingenuity called forth for overcoming human limitations. In these directions through the ever-increasing ingenuity of device for advancing science, I see at no distant future an advance of skill and of invention among our workers; and if this skill be assured, practical applications will not fail to follow in many fields of human activity.

The advance of science is the principal object of this Institute and also the diffusion of knowledge. We are here in the largest of all the many chambers of this House of Knowledge—its Lecture Room. In adding this feature, and on a scale hitherto unprecedented in a Research Institute, I have sought permanently to associate the advancement of knowledge with the widest possible civic and public diffusion of it; and this without any academic limitations, henceforth to all races and languages, to both men and women alike, and for all time coming.

[Pg_210] The lectures given here will not be mere repetitions of second-hand knowledge. They will announce to an audience of some fifteen hundred people, the new discoveries made here, which will be demonstrated for the first time before the public. We shall thus maintain continuously the highest aim of a great Seat of Learning by taking active part in the *advancement* and diffusion of knowledge. Through the regular publication of the Transactions of the Institute, these Indian contributions will reach the whole world. The discoveries made will thus become public property. No patents will ever be taken. The spirit of our national culture demands that we should for ever be free from the desecration of utilising knowledge for personal gain. Besides the regular staff there will be a selected number of scholars, who by their work have shown special aptitude, and who would devote their whole life to the pursuit of research. They will require personal training and their number must necessarily be limited. But it is not the quantity but quality that is of essential importance.

It is my further wish, that as far as the limited accommodation would permit, the facilities of this Institute should be available to workers from all countries. In this I am attempting to carry out the traditions of my country, which so far back as twenty-five centuries ago, welcomed all scholars [Pg_211] from different parts of the world, within the precincts of its ancient seats of learning, at Nalanda and at Taxilla.

THE SURGE OF LIFE

With this widened outlook, we shall not only maintain the highest traditions of the past but also serve the world in nobler ways. We shall be at one with it in feeling the common surgings of life, the common love for the good, the true and the beautiful. In this Institute, this Study and Garden of Life, the claim of art has not been forgotten, for the artist has been working with us, from foundation to pinnacle, and from floor to ceiling of this very Hall. And beyond that arch the Laboratory merges imperceptibly into the garden, which is the true laboratory for the study of Life. There the creepers, the plants and the trees are played upon by their natural environments,—sunlight and wind, and the chill at midnight under the vault of starry space. There are other surroundings also, where they will be subjected to chromatic action of different lights, to invisible rays, to electrified ground or thunder-charged atmosphere. Everywhere they will transcribe in their own script the history of their experience. From this lofty point of observation, sheltered by the trees, the student will watch this panorama of life. Isolated from all distractions, he [Pg_212] will learn to attune himself with Nature; the obscuring veil will be lifted and he will gradually come to see how community throughout the great ocean of life outweighs apparent dissimilarity. Out of discord he will realise the great harmony.

THE OUTLOOK

These are the dreams that wove a network round my wakeful life for many years past. The outlook is endless, for the goal is at infinity. The realisation cannot be through one life or one fortune but through the co-operation of many lives and many fortunes. The possibility of a fuller expansion will depend on very large endowments. But a beginning must be made, and this is the genesis of the foundation of this Institute. I came with nothing and shall return as I came; if something is accomplished in the interval, that would indeed be a privilege. What I have I will offer, and one who had shared with me the struggles and hardships that had to be faced, has wished to bequeath all that is hers for the same object. In all my struggling efforts I have not been altogether solitary while the world doubted, there had been a few, now in the City of Silence, who never wavered in their trust.

Till a few weeks ago it seemed that I shall have to look to the future for securing the necessary expansion of scope and for permanence of the [Pg_213] Institute. But response is being awakened in answer to the need. The Government have most generously intimated their desire to sanction grants towards placing the Institute on a permanent basis the extent of which will be proportionate to the public interest in this national undertaking. Out of many who would feel an interest in securing adequate Endowment, the very first donations have come from two of the merchant princes of Bombay, to whom I had been personally unknown.

A note that touched me deeply came from some girl students of the Western Province, enclosing their little contribution "for the service of our common motherland." It is only the instinctive mother-heart that can truly realise the bond that draws together the nurselings of the common homeland. There can be no real misgiving for the future when at the country's call man offers the strength of his life and woman her active devotion, she most of all, who has the greater insight and larger faith because of the life of austerity and self-abnegation. Even a solitary wayfarer in the Himalayas has remembered to send me message of cheer and good hope. What is it that has bridged over the distance and blotted out all differences? That I will come gradually to know; till then it will remain enshrined as a feeling. And I go forward to my appointed task, [Pg_214] undismayed by difficulties, companioned by the kind thoughts of my well-wishers, both far and near.

INDIA'S SPECIAL APTITUDES IN CONTRIBUTION TO SCIENCE

The excessive specialisation of modern science in the West has led to the danger of losing sight of the fundamental fact that there can be but one truth, one science which includes all the branches of knowledge. How chaotic appear the happenings in Nature? Is nature a Cosmos! in which the human mind is some day to realise the uniform march of sequence, order and law? India through her habit of mind is peculiarly fitted to realise the idea of unity, and to see in the phenomenal world an orderly universe. This trend of thought led me unconsciously to the dividing frontiers of different sciences and shaped the course of my work in its constant alternations between the theoretical and the practical, from the investigation of the inorganic world to that of organised life and its multifarious activities of growth, of movement, and even of sensation. On looking over a hundred and fifty different lines of investigations carried on during the last twenty-three years, I now discover in them a natural sequence. The study of Electric Waves led to the devising of methods for the production of the shortest electric waves known and these bridged [Pg_215] over the gulf between visible and invisible light; from this followed accurate investigation on the optical properties of invisible waves, the determination of the refractive powers of various opaque substances, the discovery of effect of air film on total reflection and the polarising properties of strained rocks and of electric tourmalines. The invention of a new type of self-recovering electric receiver made of galena was the fore-runner of application of crystal detectors for extending the range of wireless signals. In physical chemistry the detection of molecular change in matter under electric stimulation, led to a new theory of photographic action. The fruitful theory of stereochemistry was strengthened by the production of two kinds of artificial molecules, which like the two kinds of sugar, rotated the polarised electric wave either to the right or to the left. Again the 'fatigue' of my receivers led to the discovery of universal sensitiveness inherent in matter as shown by its electric response. It was next possible to study this response in its modification under changing environment, of which its exaltation under stimulants and its abolition under poisons are among the most astonishing outward manifestations. And as a single example of the many applications of this fruitful discovery, the characteristics of an artificial retina gave a clue to the unexpected discovery [Pg_216] of "binocular alternation of vision" in man;—each eye thus supplements its fellow by turns, instead of acting as a continuously yoked pair, as hitherto believed.

PLANT LIFE AND ANIMAL LIFE

In natural sequence to the investigations of the response in 'inorganic' matter, has followed a prolonged study of the activities of plant-life as compared with the corresponding functioning of animal life. But since plants for the most part seem motionless and passive, and are indeed limited in their range of movement, special apparatus of extreme delicacy had to be invented, which should magnify the tremor of excitation and also measure the perception period of a plant to a thousandth part of a second. Ultramicroscopic movements were measured and recorded; the length measured being often smaller than a fraction of a single wavelength of light. The secret of plant life was thus for the first time revealed by the autographs of the plant itself. This evidence of the plant's own script removed the long-standing error which divided the vegetable world into sensitive and insensitive. The remarkable performance of the Praying Palm Tree of Faridpore, which bows, as if to prostrate itself, every evening, is only one of the latest instances which show that the supposed insensibility of plants [Pg_217] and still more of rigid tree is to be ascribed to wrong theory and defective observation. My investigations show that all plants, even the trees, are fully alive to changes of environment; they respond visibly to all stimuli, even to the slight fluctuations of light caused by a drifting cloud. This series of investigations has completely established the fundamental identity of life-reactions in plant and animal, as seen in a similar periodic insensibility in both, corresponding to what we call sleep; as seen in the death-spasm, which takes place in the plant as in the animal. This unity in organic life is also exhibited in that spontaneous pulsation which in the animal is heart-beat; it appears in the identical effects of stimulants, anaesthetics and of poisons in vegetable and animal tissues. This physiological identity in the effect of drugs is regarded by leading physicians as of great significance in the scientific advance of Medicine; since here we have a means of testing the effect of drugs under conditions far simpler than those presented by the patient far subtler too, as well as more humane than those of experiments on animals.

Growth of plants and its variations under different treatment is instantly recorded by my Crescograph. Authorities expect this method of investigation will advance practical agriculture; since for the first time we are able to analyse and study separately the [Pg_218] conditions which modify the rate of growth. Experiments which would have taken months and their results vitiated by unknown changes, can now be carried out in a few minutes.

Returning to pure science, no phenomena in plant life are so extremely varied or have yet been more incapable of generalisation than the "tropic" movements, such as the twining of tendrils, the heliotropic movements of some towards and of others away from light, and the opposite geotropic movements of the root and shoot, in the direction of gravitation or away from it. My latest investigations recently communicated to the Royal Society have established a single fundamental reaction which underlies all these effects so extremely diverse.

Finally, I may say a word of that other new and unexpected chapter which is opening out from my demonstration of nervous impulse in plants. The speed with which the nervous impulse courses through the plant has been determined; its nervous excitability and the variation of that excitability have likewise been measured. The nervous impulse in plant and in man is found exalted or inhibited under identical conditions. We may even follow this parallelism in what may seem extreme cases. A plant carefully protected under glass from outside shocks, looks sleek and flourishing; but its higher nervous function is then found to [Pg_219] be atrophied. But when a succession of blows is rained on this effect and bloated specimen, the shocks themselves create nervous channels and arouse anew the deteriorated nature. And is it not shocks of adversity, and not cotton-wool protection, that evolve true manhood?

A question long perplexing physiologists and psychologists alike is that concerned with the great mystery that underlies memory. But now through certain experiments I have carried out, it is possible to trace "memory impressions" backwards even in inorganic matter, such latent impressions being capable of subsequent revival. Again the tone of our sensation is determined by the intensity of nervous excitation that reaches the central perceiving organ. It would theoretically be possible to change the tone or quality of our sensation, if means could be discovered by which the nervous impulse would become modified during transit. Investigation on nervous impulse in plants has led to the discovery of a controlling method, which was found equally effective in regard to the nervous impulse in animal.

Thus the lines of physics, of physiology and of psychology converge and meet. And here will assemble those who would seek oneness amidst the manifold. Here it is that the genius of India should find its true blossoming.

The thrill in matter, the throb of life, the pulse of [Pg_220] growth, the impulse coursing through the nerve and the resulting sensations, how diverse are these and yet how unified! How strange it is that the tremor of excitation in nervous matter should not merely be transmitted but transmuted and reflected like the image on a mirror, from a different plane of life, in sensation and in affection, in thought and in emotion. Of these which is more real, the material body or the image which is independent of it? Which of these is undecaying, and which of these is beyond the reach of death?

It was a woman in the Vedic times, who when asked to take her choice of the wealth that would be hers for the asking, inquired whether that would win for her deathlessness. What would she do with it, if it did not raise her above death? This has always been the cry of the soul of India, not for addition of material bondage, but to work out through struggle her self-chosen destiny and win immortality. Many a nation had risen in the past and won the empire of the world. A few buried fragments are all that remain as memorials of the great dynasties that wielded the temporal power. There is, however, another element which find its incarnation in matter, yet transcends its transmutation and apparent destruction: that is the burning flame born of thought which has been handed down through fleeting generations.

[Pg_221] Not in matter, but in thought, not in possessions or even in attainments but in ideals, are to be found the seed of immortality. Not through material acquisition but in generous diffusion of ideas and ideals can the true empire of humanity be established. Thus to Asoka to whom belonged this vast empire, bounded by the inviolate seas, after he had tried to ransom the world by giving away to the utmost, there came a time when he had nothing more to give, except one half of an *Amlaki* fruit. This was his last possession and anguished cry was that since he had nothing more to give, let the half of the *Amlaki* be accepted as his final gift.

Asoka's emblem of the *Amlaki* will be seen on the cornices of the Institute, and towering above all is the symbol of the thunderbolt. It was the Rishi Dadhichi, the pure and blameless, who offered his life that the divine weapon, the thunderbolt, might be fashioned out of his bones to smite evil and exalt righteousness. It is but half of the *Amlaki* that we can offer now. But the past shall be reborn in a yet nobler future. We stand here to-day and resume work to-morrow so that by the efforts of our lives and our unshaken faith in the future we may all help to build the greater India yet to be.

THE PRAYING PALM OF FARIDPUR

Under the presidency of Lord Ronaldshay Sir J. C. Bose delivered a lecture on Friday the 4th January 1918, at the "Bose Institute" on 'The Praying Palm-tree.' He said:

Perhaps no phenomenon is so remarkable and shrouded with greater mystery as the performances of a particular palm tree near Faridpore. In the evening while the temple bells ring calling upon people to prayer, this tree bows down as if prostrate itself. It erects its head again in the morning, and this process is repeated every day during the year. This extraordinary phenomenon has been regarded as miraculous, and pilgrims have been attracted in great numbers. It is alleged that offerings made to the tree, that is to say to the custodian of the tree, have been the means effecting marvellous cures. It is not necessary to pronounce any opinion on the subject; these cures may be taken as effective as other faith cures now so fashionable in the West.

I first obtained photographs of the two positions which proved the phenomenon to be real. The [Pg_223] next thing was to devise special apparatus to record continuously the movement of the tree day and night. But difficulties were encountered in getting the consent of the proprietor to attach foreign instruments to the sacred tree. His misgivings were however removed when it was explained that the instruments were pure Swadeshi, being made in my Laboratory. The records of the Palm Tree showed that it fell with the rise of temperature, and rose with the fall. Records obtained with other trees brought out the extraordinary and unsuspected fact that all trees are moving—such movements being in response to changes in their environment.

SENSITIVE OR INSENSITIVE?

That not a "Mimosa" alone, but all plants are sensitive was demonstrated by some striking experiments. A spiral tendril, under electric shock was shown to writhe imitating the contortions of a tortured worm. In ordinary plants, all sides being equally sensitive contraction takes place on all directions with resulting neutral effect. Another striking experiment was to show how ordinary plants could be made sensitive by the mere process of amputation of the balancing half? Further experiments were shown demonstrating the effects of light, of warmth and other stimuli on the plant. Warmth worked antagonistically to light. The [Pg_224] numerous permutations brought about by two changing variations were shown by a mechanical hand, which traced most complicated curves. In actual life the number of changing factors are very numerous, hence the intricacy involved in the manifestations of life.

The experiments that have been shown will help the audience to realise in some measure that the world we live in is not a theatre of caprice or chance, but that an all pervading law holds and regulates its destiny. We have seen that the vast expanse of life which is unvoiced, seemingly, so impassive, is instinct with sensibility. Thus the whole of the vegetable world, including rigid trees perceive the changes in their environment and respond to them by unmistakable signals. They thrill under light and become depressed by darkness; the warmth of summer and frost of winter, drought and rain, these and many other happenings leave a subtle impression on the life of the plant. By invention of apparatus of extreme delicacy, it is possible to make the plant itself write down the history of its own experience in a hieroglyphic which it is possible to decipher. From these pages, taken from the diary of the plant, it will perhaps be possible some day to get an insight into the great mystery that surrounds life itself. For I shall in the course of lectures given here show how the life of plants is [Pg_225] a mere reflection of our own. I shall show how shocks and wounds affect them as they affect animals; how a common death-throb marks the crisis when life passes into death. The exuberance of life, on the other hand, will be shown by pulsing throbs of animal's heart and spontaneous beat in vegetal tissues. Another aspect of this exuberance will be shown in the imperceptible growth of plants. My recently invented Crescograph, to be exhibited at my lecture a fortnight hence, will magnify growth a million-fold and record ultra microscopic movements, smaller than a single wave length of light. By this apparatus growth will be instantaneously recorded and conditions which foster or inhibit growth discriminated. I shall demonstrate my discovery of the nervous system in plants, and show how shocks from without pass within, and how this nervous impulse modified during transit. It will further be shown how various stimulants, anesthetics and poison induce effects which are identical in man and in plant. It will be obvious how these studies will open new fields of inquiry in different branches of science; in Physiology and Psychology; in Medicine and in Agriculture.

—*Amrita Bazar Patrika*, 7-1-1918.

VISUALISATION OF GROWTH

Sir J. C. Bose delivered on the 18th January 1918, at the Bose Institute, the second of the series of discourses on revelations of plant life. This time the audience had the opportunity of witnessing the working of Bose's newly perfected Crescograph which is undoubtedly one of the marvels in modern Science. For this apparatus gives a visual demonstration of movements which are far beyond the highest powers of microscope. The invisible internal workings of life are thus for the first time revealed to man.

LAW VERSUS CAPRICE

The lecturer first described the infinite variations in life reactions in plants. The same external stimulus, he said apparently produces one effect in one plant; and precisely opposite in another. Some leaves move towards light; others are repelled by it. The root bends towards the centre of the earth, the shoot rises above away from it. Numerous other "tropic" movements are caused by [Pg_227] contact, by electricity, by moisture and by invisible radiations. These effects appear so extremely diverse and capricious that some of the leading physiologists were forced to come to the conclusion that there was no law guiding such movement, but that the plant decides for itself what should be the effect of external conditions on it.

RECORD OF GROWTH

Most of these tropic movements are brought about by changes induced in growth by the action of different forces. But growth is so excessively slow that slight changes induced in it is impossible of detection. The proverbially slow paced snail moves two thousand times faster than the growing point of a plant. Hence to visualise growth and its changes, apparatus has to be invented which would magnify growth something like a million times. If such a thing were possible the pace of the snail would be quickened to the speed of a rifle bullet. The difficulties in connection with the devising and construction of apparatus with this extraordinary power appeared at first an impossibility. The Jewels for the fittings of the apparatus could not be found fine enough. The lecturer had to discard ordinary jewels for diamonds, such bearings being only made in Germany. But the outbreak of the [Pg_228] war put an end to this source of supply. He had then to turn to resources available in India.

ADVANCE OF AGRICULTURE

The invention of method for immediate record of growth and its variations under various conditions is one of immense practical importance. Experiments on gigantic scales are in progress all over the world for this purpose. At Rothamstead, this work has been going on for more than half a century. The great Department of Agriculture in Mashington spends millions every year on such experiments, there being a thousand men employed in research. Recently many experiments have been undertaken on the effect of electricity on growth. The results obtained have been mostly contradictory. For real advance in agriculture we must first discover the laws of growth. Ordinary experiments on growth are of little value because they take weeks for detecting changes of growth which might have been brought about by charges in the environment. The only satisfactory method is to devise an apparatus which would make the plant itself record the rate of its growth, and the changes induced by food or treatment in the course of less than a minute, during which short time it is possible to maintain external conditions constant.

THE MAGNETIC CRESCOGRAPH

All the difficulties connected with the devising of apparatus has been completely removed by the lecturer's successful invention of his new magnetic crescograph in which practically unlimited magnification is obtained without the difficulties arising from the unavoidable friction of bearings. Magnetic forces are so exactly balanced that a disturbance in the balance caused by slightest movements such as that of growth is magnified ten millions of times. The application of this new principle will be of great importance in various investigations in Physics.

Sir J. C. Bose next demonstrated some marvellous results obtained with his apparatus. A seedling which on account of the Winter season appeared stationary jotted down by taps on a moving plate, the rate of its growth. The application of a chemical instantly arrested this growth, but an antidote timely applied, not only removed the torpor but enhanced the growth at an enormous rate. The life of the plant became pliant at the will of the experimenter, and nothing appeared more marvellous than the realisation that man has the power to pierce the veil that shrouds the mystery that had hitherto baffled him.

The lecturer explained how the effect of a given agent—a chemical solution or an electric current— [Pg_230] is profoundly modified by the dose a given intensity, producing one effect and a different intensity giving rise to an effect diametrically opposite. This is the reason of the inexplicable anomalies which have baffled many investigators. Numerous are the forces which act on growth some helping, others retarding, the effects being further modified by the strength and duration of application. These factors that determine growth are each to be studied in detail, and the laws of effect of each to be discovered. There can be no real advance in scientific agriculture until this is done.

—*Amrita Bazar Patrika*, 19-1-1918.

SIR J. C. BOSE AT BOMBAY.

There was a brilliant gathering at the Royal Opera House on Tuesday the 22nd January 1918, when Sir Jagadis Bose gave a deeply interesting lecture on the history of the inception of his Institute in Calcutta and its aims together with an exposition of his scientific researches illustrated by lantern slides. The theatre was full long before the lecture commenced and several prominent people were present the bulk of the audience consisting of Indians.

Mr. Tilak in introducing the distinguished lecturer to the audience referred to Professor Bose's lasting services not only to the Indian nation but to the whole world. These references to Dr. Bose and his work elicited frequent applause from the large audience.

A FIFTY THOUSAND RUPEES LECTURE.

Sir Jagadis, who was accorded a most enthusiastic ovation on rising to address the gathering, acknowledged his gratitude to the public of Bombay who proved their appreciation of his work by their [Pg_232] presence there that evening, and the fact that they had subscribed Rs. 50,000 for the occasion. He then gave a brief explanatory account of the nature and scope of his work, which he had planned and carried out alone for many years amidst many and varied difficulties. He gave an exposition by the aid of one of the delicate instruments of his own invention of how plants respond to various sounds and tunes and the beautiful colour display which was observed in this connection appeared as though he were a magician with a wand.

PLANTS UNDER ANAESTHETICS

The Doctor explained the meaning and significance of the thunderbolt which has been adopted as the symbol of the institution. He explained also the special uses to which the various parts of the buildings would be put. The fact was brought out that the entire building and grounds had been designed to suit the special needs of the Institute and care had been taken to make it as far as possible self contained. An interesting feature of the garden close to that portion which forms the residence of Sir Jagadis was the open platform perched above two trees, transplanted under anaesthetic conditions. A variety of apparatus is displayed under these trees and the platform is intended for observation or meditation or both. Dr. Bose here explained [Pg_233] how trees when transplanted frequently died under the shock of the operation just as human being sometimes died, not from an operation but from the shock caused thereby. Similarly he had discovered and proved that trees could, like human beings, go through severe operations and survive the shock, if placed under the influence of an anaesthetic.

SOME PHENOMENA OF PLANT LIFE

The Professor explained next other experiments which he had performed on plants and whose results had exhibited the close parallel which plant life bears to human life. With the aid of another delicate instrument he showed how the growth of plants can be influenced by drugs and the demonstration on the screen of the manner in which the slow growth of a plant can be thus expedited was one of extraordinary interest. One was able to see the flame of life moving up the screen and recording at intervals the stages of growth, a lengthening of the intervals between each recorded glow illustrating the acceleration of growth as soon as the drug was applied. The instruments necessary to record this phenomenon are of extraordinary delicacy, and barely survived the strain of the journey from Calcutta.

ELECTRICITY [Pg_234] AND AGRICULTURE

The last experiment was in regard to the effect of electricity on plant life. He referred particularly to the fact that it was his aim to discover the law of growth and atrophy among plants. Such a discovery had a great bearing on the future of agriculture and would revolutionise world thought. Electricity, he explained and illustrated, would promote or retard the growth of life by reaction. In England and other countries electricity had been applied to agriculture but without exact knowledge of its varying effect on plant life. He then showed by another apparatus of extreme delicacy that electricity might retard and even repel as well as promote the growth of plant life. But if the law of growth and decay could be ascertained, it was possible to regulate the control of life under most varied conditions.

—*Amrita Bazar Patrika*, 29-1-1918.

UNITY OF LIFE

Under the auspices of the Bombay University, Sir Jagadis Chundar Bose delivered on Thursday, the 31st January 1918, a lecture on the "Unity of Life." It was illustrated by lantern slides and an instructive exposition was given of some of his unique discoveries in the realm of Plant Life....

HIDDEN HISTORY IN PLANTS LIFE

"The subject of my address to-night is the 'Unity of Life.' Under a placid exterior there is a hidden history on the life of the plant. Is it possible to make the plants write down their own autographs and thus reveal their history? In order to succeed in this we have first to discover some compulsive force which will make the plant give an answering signal, secondly, we have to invent some instrument of extreme delicacy for the automatic conversion of these signals into an intelligent script; and last of all, we have ourselves to learn the nature of the hieroglyphics."

Sir J. C. Bose then explained the principle of his epoch-making Resonant Recorder which writes [Pg_236] down the perception period of the plant within a thousandth part of a second, and writes down the action of light and warmth and drugs on the plant; the effect of vitiated air, of passing clouds, of excess of food and of drink.

"The plant is very human in its virtues and weakness. Plants like animals become exalted, grow tired or despond. An easy greenhouse life makes them less than themselves, overgrown and flabby, capable of response, till they have become hardened by adversity to a fuller existence. A time comes when after an answer to a supreme shock, there is a sudden end of the plant's power to give any further response. This supreme shock is the shock of death. Even in this crisis there is no immediate change in the placid appearance of the plant. Drooping and withering are events that occur long after death itself. How does the plant then give its last answer? In man at the critical moment a spasm passes through the whole body and similarly in the plant I find a great contractile spasm takes place. This is accompanied by an electrical spasm also. In the script of the Death Recorder the line that up to this time was being drawn, become suddenly reversed and then ends. This is the last answer of the plant.

"These our mute companions, silently growing beside our door, have now told us the tale of their [Pg_237] life-tremulousness and their death-spasm in script that is as inarticulate as they. May it not be said that this story has a pathos of its own beyond any that we may have conceived?

"We have now before our mind's eye the whole organism of the perceiving, throbbing and responding plant, a complex unity and not a congeries of unrelated parts. The barriers which separated kindred phenomena in the plant and animal are now thrown down. Thus community throughout the great ocean of life is seen to outweigh apparent dissimilarity Diversity is swallowed up in unity.

"In realising this, is our sense of final mystery of things deepened or lessened? Is our sense of wonder diminished when we realise in the infinite expanse of life that is silent and voiceless the foreshadowings of more wonderful complexities? Is it not rather that science evokes in us a deeper sense of awe? Does not each of her new advances gain for us a step in that stairway of rock which all must climb who desire to look from the mountain tops of the spirit upon the promised land of truth?"

Sir Jagadis then gave a most interesting exposition of his researches with the aid of magic lantern slides.

SENSITIVENESS IN PLANTS

Referring first of all his discovery of sensitiveness [Pg_238] in plants, he said that in that respect they were akin to the human system. He illustrated this truth by a demonstration of the reaction that takes place in the frog when a shock is communicated and side by side presenting the reaction that is similarly effected in the plant. "Plants have a nervous system like our own," he said, and with the aid of an enlarged illustration of the mimosa he showed the changes that took place when the plant was disturbed. Turning to plant autograph, he spoke of the Resonant Recorder, a special apparatus which he has invented to prove how even plants are tuned to environment. Certain tunes had no effect on plants, he said, while others had and he asked them specially to observe the beautiful and variegated colour formation produced by their response to tunes. He gave an interesting experiment on this point, and both Lord and Lady Willingdon tried it. There was a great outburst of cheering, which was renewed each time the effect was produced, and it was noticed that the cheering, which was vociferous had its own effect. It had taken him a long time, he said, to produce and perfect the complete apparatus to determine the latent mimosa and by the aid of that apparatus, he was able to record the movement of the plant to one thousandth of a second.

He next went on to say that all plants were [Pg_239] endowed like ourselves, but at first the news was received with great scepticism. He did not despair, however, of success and was continuously engaged in discovering, in collecting fresh evidence. Thanks to the action of the Government of India in sending him on a world tour, he got at last the opportunity to prove before the scientific societies of the world, the truth of his discoveries. An illustration of the Mimosa which has accompanied him in his world tour was screened.

The next illustration was to show how long plants took to feel shock and what time they took to recover. Like the great human system plants were subject to periodic conscianimal [sic., consciousness?] had their periods of sleep and awakening. The extra water pressure produced during sunset had nothing to do with true sleep. Plants, too, were subject to exaltation and depression and at certain hours of the day they were fully conscious and active while at other hours they were dormant and lazy. He showed by means of a chart that they were fast asleep between 6 and 9 in the morning and his humorous remark that in that respect they had taken a leaf from our modern society ladies provoked a great deal of laughter. A series of records were then shown to illustrate the various degrees of plant consciousness, which were deeply appreciated by the audience.

Proceeding Dr. Bose said that plants were far [Pg_240] more conscious of nature than human beings and described his experience how plants were sensitive even to passing clouds, which produced on them a depressing effect. He spoke of the difference between thin and wiry grown plants and those that were stout and robust. In that respect they resembled again human beings and thin and wiry grown plants were far more susceptible of excitement than the others. They, too, needed rest and without it, they were flabby and depressed. A cartoon from the London "Punch" entitled "A successful Trial" was screened to the merriment of the audience, in which the Professor was humorously depicted by that journal, after his exposition before the Royal Institute in London. He gave an illustration of the "Praying Palm of Faridpur" and the changes it exhibited to environment. All plants displayed similar power and these changes were no longer inscrutable. They had been brought within the realm of scrutability [sic.] and could be recorded.

"PROTECTING" PLANTS

It was a mistake to suppose that when "protected" plants would thrive better. Mothers had a tendency to keep their children away from contact with the outside world with a view to "protect" them. He had placed a plant under a glass case and the effect of it was he had a gloated and effete [Pg_241] specimen, flabby-looking in appearance and weary under adversity, they recovered sooner and their growth was healthy just as it evolved true manhood in men. It had been commonly believed that carbonic acid gas was conducive to plant growth. That was a great mistake. In sunshine, plants readily absorbed it; but it was no more true that plants thrived on CO_2, than did human beings. He illustrated the effect of carbonic acid gas as well as oxygen. The latter was as much necessary for plants to thrive on as it was for them. Another illustration exhibited the effect of alcohol on plants and he declared amidst laughter that alcohol produced the same alternate maudlin depression and exaltation on plants that is to be observed on the human system. He said that this experiment had tickled the Americans a great deal and referred to a conversation he had with Mr. Bryan, who was a teetotaller, regarding alcohol given to plants. Some American papers had given characteristic headlines to introduce his lecture on the effect of stimulus to plants.

Another plant Desmodium which has accompanied him in his world tour was filmed on the screen. He spoke, next, of the apparatus which he had invented to record plant pulsation and the struggle they exhibited between life and death. Poisons had as much effect on plants as on men, and they [Pg_242] could be revived by applying antidotes, this was illustrated by another chart. Another point of interest dealt with by him was the effect of warm water on plants, and he gave an exposition of his discovery to show that plants died when placed in 60 degree (centigrade) warm water. He referred to the stupendous phenomenon of invisible writing by means of which the plant recorded its own evolution.

The lecture was listened to with profound interest and lasted for an hour. Mr. Setalvad proposed a hearty vote of thanks to the Chancellor for presiding at the meeting. Lord Willingdon, in acknowledging it, said that the vote of thanks was due to Sir Jagadis rather than to himself. As he had anticipated in the beginning, the lecture had proved absorbingly interesting and he was afraid Sir Jagadis's discoveries might be positively alarming when he next visited Bombay. He hoped that they would accord Sir Jagadis a hearty vote of thanks with "true Bombay cordiality." After a few suitable remarks by Sir Jagadis the meeting terminated.

—*Amrita Bazar Patrika*, 5-2-1918.

THE AUTOMATIC WRITING OF THE PLANT

On the 8th February 1918, Sir J. C. Bose delivered the following discourse on 'The Automatic Writing of the Plant,' at the Bose institute: —

Sir J. C. Bose spoke of two different ways of gaining knowledge, the lesser way is by dwelling on superficial differences, the mental attitude which makes some say 'Thank God I am not like others:' The other way is to realise an essential unity in spite of deceptive appearance to the contrary. He had recently been on a visit to the western Presidency, he went there as a stranger, but he has come back with a pang at parting from kindreds. Never in his life did he realise so vividly as now the great unity that drew together all who regarded India as their home and place of work. They were bound to each other by mutual ties of dependence. He had for many years been engaged in discovering community in physical manifestations of life. Now he has realised an abiding unity in the highest manifestations of human life, in community of thoughts and ideals.

[Pg_244] In the wide expanse of life itself few things would appear so strikingly different as the life activities in plants and in animals. But if in spite of the seeming differences, it could be proved that these life activities are fundamentally similar, this would undoubtedly constitute a scientific generalisation of very great importance. It would then follow that the complex mechanism of the animal machine, that baffled us so long, need not remain inscrutable for all time, for the intricate problems of animal physiology would then naturally find their solution in the study of corresponding problems under simpler conditions of vegetative life. That would mean an enormous advance in the science of physiology, of agriculture, of medicine, and even of psychology.

How then are we to know what unseen changes take place within the plant? The only conceivable way would be, if that were possible, to detect and measure the actual response of the organism to a definite testing blow. When an animal receives an external shock it may answer in various ways; If it has voice, by a cry, if dumb, by the movement of its limbs. The external shock is the stimulus, the answer of the organism is the response. If we can make it give some tangible response to a questioning shock, then we can judge the condition of the plant by the extent of the answer. In an excitable condition the feeblest stimulus will evoke an extraordinarily [Pg_245] large response, in a depressed state even a strong stimulus evokes only a feeble response, and lastly, when death has overcome life, there is an abrupt end of the power to answer at all.

Prof. Bose then explained the principle and action of his apparatus by which the plant attached to it is automatically excited by successive stimuli which are absolutely constant. In answer to this the plant makes its own responsive records, goes through its own period of recovery, and embarks on the same cycle over again without assistance from the observer at any point. In this way the effect of changed external conditions is seen recorded in the script made by the plant itself.

It has been thought that plants like mimosa alone were sensitive. But Sir J. C. Bose's apparatus demonstrated the unsuspected fact that every plant and every organ of every plant answered to a shock by a contractile spasm, as by an animal muscle. If perception of feeble stimulus be taken as a measure of ascent in the scale of life then the superiority of man must be established on a foundation more secure than sensibility. The most sensitive organ by which we can detect electric current is our tongue. An average European can perceive a current as feeble as six micro-amperes, a micro-ampere being a millionth part of the electric unit. Possibly the tongue of a Celt is more excitable, and I have [Pg_246] no doubt that my countrymen can easily boast the Celt in this particular test. But the plant mimosa is ten times more excitable than the tongue of an advocate in this province.

Professor Bose then showed how identical were the effects of light, warmth and various drugs on the plant and animal. These experiments bring the plant much nearer than we ever thought. We find that it is not a mere mass of vegetative growth, but that its every fibre is instinct with sensibility. We are able to record the throbbings of its pulsating life, and find these wax and wane according to the life conditions of the plant, and cease in the death of the organism. In these and many other ways the life reactions in plant and man are alike, and thus through the experience of the plant, it may be possible to alleviate the sufferings of man.

—*Amrita Bazar Patrika*, 9-2-1918.

CONTROL OF NERVOUS IMPULSE

At the first anniversary meeting of the Bose institute, held on the 30th November 1918, Sir J. C. Bose gave the following discourse on his recent discoveries relating to the question of control of nervous impulse, under the Presidency of His Excellency Lord Ronaldshay, Governor of Bengal.

It is one of the greatest of all mysteries how we are put in connection with the external world: how blows from without are felt within. Our organs of sensation are like so many antennae radiating in various directions and picking up messages of many kinds. All of these, when analysed to their utmost, consist of shock effects on different chords. An extremely feeble stimulus is below the limit of perception, a moderate stimulus transmits excitation, which is perceived as sensation of not an unpleasant character, but the tone of sensation becomes painful when the excitation is very intense. Our sensation is thus coloured by the intensity of the nervous excitation that reaches the central organ. We are subject to human limitations, through the imperfection of our senses on the one hand, and [Pg_248] over-sensibility on the other. There are happenings which elude us because the impinging stimulus is too feeble to waken our senses; the external shock, on the other hand, may be so intense as to fill our life with pain.

Since we have no direct power over the shocks which come to us from the outside world, is it possible to control the nervous impulse so that it should be exalted in one case, and inhibited or obliterated in the other? Does advance of science hold any such

possibility? This question is plainly fraught with high significance.

PROBLEM OF CONTROL OF NERVOUS IMPULSE

Before proceeding further it will be necessary first to obtain a clear idea of the function of a nervous tissue and its characteristics; secondly the manner, in which the nervous impulse is propagated;

and lastly, we have to discover some compulsive force by which the impulse may be intensified or inhibited during transit. The nerve circuit may be liked to an electric circuit, and invisible impulse bringing about response in the indicator, be it the brain or the galvanometer. In the electric circuit the conducting power of the metallic wire is constant, and the intensity of the electric impulse depends on the intensity of the electric force applied. If the conducting power of [Pg_249] the nerve were constant then the intensity of the nervous impulse and its resulting sensation would depend inevitably on the intensity of the shock from outside which starts the impulse. In that case the possibility of the modification of our sensation would be an impossibility. But there may be a likelihood that the power of conduction possessed by a nerve is not constant but capable of change. Should this surmise prove to be correct then we arrive at the momentous conclusion that sensation itself is modifiable, whatever the external stimulus. For the modification of nervous impulse there remains only one alternative; namely, some power to render the vehicle a very much better conductor or a non-conductor according to particular requirements. We require the nervous path to the supra-conducting to have the impulse due to feeble stimulus brought to sensory prominence. When the external blow is too violent we would block the painful impulse by rendering the nerve a non-conductor.

Under narcotic the nerve becomes paralysed and we can by its use save ourselves from pain. But such heroic measures are to be resorted to in extreme cases, as when we are under the surgeon's knife. In actual life we are confronted with unpleasantness without notice. A telephone subscriber has an evident advantage, for he can switch [Pg_250] off the connection when the message begins to be unpleasant. Statesmen or politicians have been known to cultivate convenient deafness; but that is a mere pretence. The unpleasant things heard, would still continue to rankle. It is not every one that has the courage of Mr. Herbert Spencer who openly resorted to his ear plugs whenever his visitor became tedious.

The lecturer then explained that the propagation of nervous impulse is a phenomenon of transmission of molecular disturbance. It occurred to him that the transmission could be controlled if he succeeded in discovering a compulsive force which would confer on the conducting particles two opposite molecular dispositions, one of which would exalt and the other resist the impulse. His experiments were first conducted with the primitive type of nerve which he had previously discovered in plants. In full confirmation of his theory, he succeeded in conferring on the nervous tissue two opposite dispositions. Under favourable disposition the nerve is rendered supra-conducting; subliminal stimulus now becomes fully perceived. Under the opposite molecular disposition the violent impulse due to excessive stimulus becomes weakened or arrested during transit, and the plant remains quite unaffected by the external shock.

The lecturer has in his previous works demonstrated [Pg_251] the unity of life-reactions in the plant and animal. A climax is now reached when by the application of identical treatment he is able to confer alternately on the same animal nerve, supra-conducting or non-conducting property at will. Under a particular molecular disposition the experimental frog perceived and responded to stimulus which had hitherto been below its threshold of perception. Under the opposite disposition violent tetanic spasm caused by the irritant salt applied to the nerve became at once quelled. The normal property of the nerve was at once restored on the withdrawal of the predisposing force.

MAN VICTORIOUS OVER CIRCUMSTANCE

Thus by the control of molecular disposition of the conducting nerve, nervous impulse, and the resulting sensation may become profoundly modified. The external is not so overwhelmingly dominant, and man is not to be merely passive in the hands of destiny. There is a latent power which would raise him above the terrors of his inimical surroundings. It remains with him that the channels through which the outside world reach him should, at his command be widened or become closed. It may thus be possible for him to catch those indistinct messages that had hitherto eluded him or he may withdraw within himself, so that in [Pg_252] his inner realm, the jarring notes and the din of the world should no longer affect him.

The whole audience heard the discourse with spell bound interest. The Indian Scientist came to that realisation by experiments at which the Indian Jogis of yore arrived by intuition. Following an absolutely original line inventing his own apparatus of the most simple yet subtle delicacy and having constructed them by the hands of Indian artisans, working without collaborators and with the smallest modicum of recognition by his fellow scientists, he has pursued his investigation to a result which has been a revelation to the whole world. Dr. Bose has proved that man and plant are one body and life in their physiology, in their vital habits and nervous responses. He has clearly demonstrated that nervous life in the plant responds to the same stimuli as in human beings. He has established between animal and plant a unity of incipient mind. The plant not only lives and dies, wakes and sleeps but it makes the responses which in animal would be pleasure and pain.

Dr. Bose has made a great step towards the unification of knowledge. A bridge has been built between man and inert matter. Even if we take Dr. Bose's experiments with metals in conjunctions with his experiments on plants, we may hold it to be practically proved for the thinker that Life in [Pg_253] various degrees of manifestation and organisation is omnipresent in Matter and is no foreign introduction or accidental development, but was always that to be evolved.

The ancient thinkers knew well that life and mind exist everywhere in essence and vary only by the degree and manner of their emergencies and functionings. All is in all and it is out of complete involution that the complete evolution progressively appears. It is only appropriate that for a descendant of the race of ancient thinkers who formulated that knowledge, should be reserved the privilege of initiating one of the most important among the many discoveries by which experimental science is confirming the wisdom of his forefathers.

—*Amrita Bazar Patrika*, 4-12-1918.

MARVELS OF GROWTH AS REVEALED BY THE "MAGNETIC CRESCOGRAPH"

[Sir J. C. Bose has recently invented the "Magnetic" crescograph. It is a supersensitive instrument and the very high magnification obtained by it surpasses all existing appliances. By this instrument, phenomena hitherto beyond the reach of investigation can now be studied with great precision. It shows ultra-microscopic changes inducted in a growing organism even by a puff of smoke or a gentle breeze, by a passing cloud or fleeting brightness. This super magnifier was exhibited for the first time by Sir J. C. Bose before an appreciative gathering 10-1-1919. A number of lady students, professors, lawyers, doctors and several eminent personages gathered to hear the great Indian scientist.]

In his Discourse on the above subject on Friday, Sir J. C. Bose illustrated how the limitations imposed on the advance of science by the imperfection of our senses, may stimulate the invention of supersensitive apparatus which reveals to us the existence of phenomena hitherto unknown. Thus [Pg_255] the invention of the microscope from a simple lens magnifying 3 or 4 times into progress up to 1500 diameters has given birth to new sciences. But still higher magnification is demanded in unravelling the mystery of movements associated with the simplest type of life as seen in plants. Greatest potentiality in life is often latent; the gigantic banian tree grows out of a thing which is smaller than the mustard seed. Within the seed-coat the dormant life remains in safety, protected from dangers outside. The seeds may thus be subjected without harm to cold so intense as will freeze mercury into solid and air into liquid. Winds and hurricanes scatter the seed of life and the cocoa-nut rides the tumultuous waves till anchored safe in an island yet to be inhabited. In due season there begins a series of most astonishing transformations; the latent life wakens, and the seedling begins to grow. The root turns downwards and the shoot upwards. Underground, the root winds its way round stones and obstacles towards moist places. Above ground the stem bends as if in search of light. Tendrils twine about a support. These visible movements are striking enough, but within the unruffled exterior of the plant body there are others, energetic and incessant, which escape our scrutiny. The bending of a growing organ towards or away from stimulus must be due to unequal growth on [Pg_256] two sides of the organ, a retardation of growth on the proximal or acceleration on the distant sides. Various theories have been advanced which have proved inadequate. For the identical stimulus of gravity produces one kind of curvature in the root and the very opposite in the shoot. The possibility of direct experimental investigation has been frustrated by the excessive slow rate of growth rendering accurate measurement impossible.

THE SLOWNESS OF GROWTH

The movement of growth is two thousand times less rapid than the place of the proverbially slow-footed snail. Taking the average annual growth in height of a tree to be 5 ft., it will take a tree a thousand years to cover a distance of a mile. We take a piece of 2 ft. in the course of half a second, during the interval plant grows through a length of 1,100,000 part of an inch or half the length of a wave of light. For investigation on the effect of external conditions on growth we have to measure even a fraction of that excessively small length.

The peasant has eagerly watched the growth of his plants on which his own life and the world's depend and, even realised something of its vicissitudes, so the vegetable physiologist has here one of the many problems of his science. The invention of growth-measuring instruments has thus been one [Pg_257] of his main endeavours. He has hitherto succeeded by the use of levers with unequal arms to obtain a magnification of about 20 times, and even then it takes many hours for growth to become perceptible; owing to the practical impossibility of maintaining the external conditions constant for so many hours, the results of measurement of growth become vitiated. It is therefore necessary to produce a magnification so high that growth should become measurable in less than a minute. The first improvement effected by the lecturer, now some fourteen years ago, was his Optical Lever, which at once raised the magnification from 20 to 1000 times, an advance which at the time seemed to many incredible, but it is at length coming into use in advanced laboratories in Europe.

THE RECORDING CRESCOGRAPH

A new apparatus devised by the lecturer, the Recording Crescograph, is described in the Transactions of the Royal Society, and of the Bose Institute. By a compound system of levers the magnification is raised to 10,000 but this is not without great technical difficulties, which cost five years of efforts to overcome. Thus the levers require to be extremely light; this was secured by the use of an alloy of aluminium used in the construction of Zeppelins: this combines lightness with rigidity. Another difficulty almost unsuperable [Pg_258] arises from the friction at the bearings of the fulcrum, the best watch jewels made of ruby were employed, but the supply was cut off from Germany by the war. This proved a blessing in disguise, for it forced the lecturer to devise a new principle of suspension using local material. This was found in practice to be far superior to jewel bearings, which became clogged by invisible dust particles present in the air. With this Recording Crescograph many phenomena of extreme interest have been discovered. The plant itself not only recorded its normal rate of growth but the slightest change induced in it by the action of different forces. So delicate was the apparatus that it analysed growth into a series of pulses, a sudden shooting out followed by a partial recoil. It showed how the growth of the plant was retarded by a mere touch, and the time it took the plant to recover from the effect of contact, and all these in course of a few seconds. The effect of different food on growth, the effect of different drugs, or living capacity these and many more became revealed by the automatic record made by the plant. This has opened out fresh and more exact method of medical inquiry, and of practical agriculture.

THE MAGNETIC CRESCOGRAPH

Such unlooked for results called for yet higher [Pg_259] magnification, and at first it seemed that further multiplying lever might be added to the previous system. But this failed on account of added mass and friction; and some altogether new solution had therefore to be sought. Material contact having proved unworkable the ideal weightless and frictionless linking was obtained by introducing a new magnetic contrivance, and this with the surprising potency of magnification from 5 to 100 million times. The mind cannot grasp the meaning of this stupendous magnification; how then could we translate it in terms which may be understood? Let us take once more our slow-footed snail, a magnification of ten million times would convert its speed to something for which there is no parallel even in modern gunnery practice. The 15 inch cannon of the "Queen Elizabeth" has a muzzle velocity of 2360 ft. per second or 8-1/2 million feet per hour. But the speed of the snail when magnified ten million times would render it 200 million ft. per hour or 24 times faster than the fastest cannon shot. We may next turn to the cosmic movement for a parallel: A point in equator whirls round at the rate of 1037 miles per hour. But a snail with the magnified speed would beat the earth by going round 40 times during the period the earth makes but one revolution!

LIFE IN STATE OF SUSPENSE AND ITS SUBSEQUENT RESOLUTION

With the experiments carried with the Magnetic Crescograph life becomes subservient to the will of the experimenter. The rate of growth is indicated by the speed with which a spot of indicating light moves across the scale. The actual rate of growth is fifty thousandth part of an inch per second; this under magnification is seen by the indicating spot of light to move at the rate of 36 inches per second: this is the normal rate. The plant is made to imbibe soda water and the growth becomes suddenly exalted some ten times; but a puff of tobacco smoke instantly retards the rate. To induce further retardation a depressing drug is next applied. The growth gradually comes to a stop and the quiescent of the spot of light shows life in a state of suspense. The plant is now hovering in an unstable poise between life and death, a slight tilt one way, and life gets interlocked in the rigidity of death. But the antidote is applied just in time, the torpor and suspense is over, and life renews her activity once more with the fullest vigour.

It is true that man is but poorly provided for his voyage of discovery in seas unknown, he can hear little and see less. A single octave of light circumscribes his vision; even of the visible the size of the ripple of light imposes an impassable barrier. [Pg_261] But he has not been deterred by his limitations but has on the contrary been spurred on its greater efforts in his explanation of the invisible. The mysterious movements of life are not to remain for him inscrutable and indecipherable for all times: but his untiring and single-minded pursuit will someday reveal to him the secret that lies behind the manifestations of life.

—*Amrita Bazar Patrika*, 13-1-1919.

THE NIGHT-WATCH OF NYMPHAEA

Sir J. C. Bose gave the following Discourse on the 'Night-Watch of Nymphaea,' at the Bose Institute, on the 24th January, 1919.

[Sir J. C. Bose's discourse delivered at the Bose Institute, on the 24th January, 1919, dealt with the mysterious phenomenon of recurrent opening and closure of flowers. Some of them open in the morning and close in the evening; others do exactly the opposite opening at night and closing during the day. These various effects have been described as the 'waking' and 'sleep' movements of plants. The subject had attracted the attention of plant physiologists for more than half a century. After summarising the various results lost in his recent work says that no satisfactory explanation of the sleep movements of plants has yet been forthcoming and that the true theory can only be established after new and exhaustive research. This investigation has been in progress at Sir J. C. Bose's laboratory for the last five years; and special automatic recorders have been invented by means of which [Pg_263] numerous plants have been recording their movements for every hour of the day and night and for many days in succession.]

In course of his discourse the lecturer said "The poets have forestalled the men of science. Why does the water-lily 'Kumud or Nymphaea' keep awake all night long and close her petals during the day? Because the water-lily is the lover of the Moon and like the human soul expanding at the touch of the beloved, the lily opens out her heart at the touch of the moon beam, and keeps watch all night long; she shrinks affrighted by the rude touch of the Sun, and closes her petals during the day. The outer floral leaves of the lily are green, and in the day time the closed flowers are hardly distinguishable from the broad green leaves which float on the water. The scene is transformed in the evening as if by magic, and myriads of glistening white flowers cover the dark water.

"The recurrent daily phenomenon has not only been observed by the poets, but an explanation offered for it. It is the moonlight then that causes the opening of the lily, and the sunlight the movement of closure. Had the poet taken out a lantern in a dark night; he would have noticed that the lily opened at night in total absence of the moon; but a poet is not expected to carry a lantern and peep out in the dark; that inordinate curiosity is characteristic [Pg_264] only of the man of science. Again the lily does not close with the appearance of the sun; for the flower often remains awake up to eleven in the forenoon. A French dictionary maker saw Cuvier, the Zoologist about the definition of the crab as 'a little red fish which walks backwards.' 'Admirable,' said Cuvier. 'But the crab is not necessarily little, nor is it red till boiled; it is not a fish, and it cannot walk backwards. But with these exceptions your definition is perfect.' And so also with the poet's description of the movement of the lily, which does not open to moonlight, nor yet close to the sun."

THE 'SLEEP' AND 'WAKING' OF JHINGA FLOWER

The waking and sleeping of the water lily is by no means an isolated instance. My attention was first drawn to another remarkable floral display by the folk song which begins with:

"Our day of work is over
Like life's span, but an hour!
For now behold the gold-started fields
Of opening 'Jhinga' flowers!"

Since then I witness every afternoon a glorious transformation in my experimental garden at Sijbaria on the Ganges. The gardener has planted a large field with Jhinga (Luffa acutangula). The [Pg_265] flowers when closed at day time are very inconspicuous, the lowest whorl of the sepals being dull green: in my afternoon walk I can hardly recognise the old familiar field, which is now covered with masses of flower in their golden glory. Here also the flowers remain open throughout the night; but they close early in the morning and the fairy field of cloth of gold vanishes suddenly.

COMPLEXITY OF THE PROBLEM

The revolutions made by the plant-scripts led to the discovery of certain new and unsuspected reactions in the life of plants, notably the influence of variation of temperature in modifying the geotropic curvature. There are at least ten variables, which by their joint effects give rise to over a thousand variations in the resulting movement of plants. The effect of each of these different factors has been isolated and a new theory propounded which offers a complete explanation of the so called sleep movements. The life reactions of plants to the various stimuli of the environment was most strikingly illustrated by means of supersensitive Magnetic Crescograph. The plant was shown to perceive the shock of light, to which it made an answering signal, so also to the action of warmth and cold. And it was explained how the various combinations of effects induced by environmental [Pg_266] change found diverse expressions in the movement of plants.

The scientific explanations offered for the opening and closing of the water lily is that the flower is closed under sunlight and that the opening takes place under darkness. But Prof. Bose has been able to keep the lily awake even in day time by placing it in a cool place. Simultaneous record of the movement of the flower and the thermograph of daily variation of temperature proved conclusively that a rapid fall of temperature in the evening brought about the opening of the flower, at first slowly then rapidly, and by 10 p.m. the flower was fully expanded. About 6 a.m. in the morning there is a rise of temperature, and the reverse movement of closure sets in. The flower continues to close very rapidly the sleep movement of closure is complete by about 10 a.m.

It will be seen how different flowers through their sensitiveness to heat and cold execute movements of "sleep" or of "waking." Some of them have the healthy habit of normal humanity to sleep at night and keep awake at day-time. Others turn night into day, and make up for their long night watch by sleeping it off at the day-time.

—*Amrita Bazar Patrika*, 25-1-1919.

WOUNDED PLANTS

Sir J. C. Bose delivered the following lecture on the 'Wounded Plants' at the Bose Institute, on the 7th February, 1919: —

It is a little over four years now that the Embodiment of World Tragedy stalked over Western Europe. The fair field of France and the bright sky was under a pall of battle-smoke. Our sight could not penetrate through the dense gloom, and the mortal cry of the wounded and dying, drowned by hoarse roar of a thousand did not reach our ear. But from the time the Sikh and the Pathan, the Gurkha and the Bengali, the Mahratta and the Rajput flung themselves in front of battle from that day our perception has become intensified. The distant cry of those whose life-blood has crimsoned the white field of snow, has found reverberating echo in our heart. What is that subtle bond by which all distances are bridged over, and by which an individual life becomes merged in larger life? Sympathy is that bond by which we come to realise the unity of all life. Before us are spread multitudinous plants, silent and seemingly impassive. They too [Pg_268] like us are actors in the Cosmic drama of life, like us the play thing of destiny. In their checkered life, light and darkness, the warmth of summer and frost of winter, drought and rain, the gentle breeze and whirling tornadoes, life and death alternate. Various shocks impinge on them, but no cry is raised in answer. I shall nevertheless try to decipher some chapters of their life history.

When a man receives a blow or shock of any kind, his answering cry makes us realise that he is hurt, but a mute makes no outcry. How do we realise his sufferings? We know it by his agonised look by the convulsive movement of his limbs, and through fellow-feeling realise his pain. When a frog is struck it does not cry, but its limbs show convulsive movement. But from this it does not follow that the frog is not hurt, for some would urge that there is a great gap between us and lower animals. One who feels for the humblest of His creatures alone knows whether the frog is hurt or not. Human sympathy always aspires: it is sometimes extended to equals, hardly ever to inferiors. And so it happens that many would doubt, whether the lowly and the depressed possess the fine sense of the exalted to feel the same joy and sorrow, and to resent social tyranny. When human attitude is so finely discriminative as regards different grades of his own species, it might be [Pg_269] extravagant to believe that the frog could have any consciousness of pain. A concession might however be made that the frog perceives a shock to which it responds by convulsive movements. It is as well that we should be careful about the use of terms for an eminent biologist insisted that animals never felt any pain: when an oyster is swallowed alive, it did not, according to him, feel any pain but rather a sensation of grateful warmth at contact with the alimentary tract. The question will remain undecided for no one has as yet returned from the gastric cavity of the tiger to expatiate on the exquisite sensation.

TEST OF LIVINGNESS

Responsive movements being a test of life, we shall try to construct a scale with which the height of livingness may be measured. What is the difference between the living and the dead? The living answers to a shock from without; the most lively gives the most energetic, the torpid or dying the feeblest, and the dead no answer at all. Thus life may be tested by shocks from without, the size of the answer being the gauge of vitality. The answer of the strong will be violent and almost explosive in its intensity, while the weakling will barely protest. The responsive movements may be recorded by suitable apparatus. The successive responses to [Pg_270] similar shocks will remain uniform, if the living tissue remained always the same. But the living organism is always in a state of change for environment is always building us anew, and we are changing everyday of our life. We are thus subject to change, some day we are in a state of high exuberance, and other time in a state of lowest depression: we pass through numerous phases between the two extremes. Not merely does the present modify, but there is also the subtle impress of memory of the past. The sum total of all these characterise one individual from another. How is the hidden to be made manifest? To test the genuineness of a coin, we strike it and the sound response betrays the true from the false. The genuine rings true and the other gives a false note. In this way perhaps the inner history of different lives may be revealed by shocks and the resulting response.

EFFECT OF WOUND

There are three separate investigations that have been carried out on the effect of wound on plants: The first is the shock effect of wound on growth: this generally speaking retards or arrests growth. In the second series of investigations the change of spontaneous pulsation of the leaflet of the Telegraph plant was recorded. Death begins to spread from the cut end of the leaflet, and reaches the throbbing [Pg_271] tissue which becomes permanently stilled on cessation of life. Experiments are in progress of arrest their march of death, and the cut leaflet which died in 24 hours has now been kept alive for more than a week.

PARALYSIS OF SENSIBILITY

Another series of investigations were carried out on the paralysing effect of severe wound. A leaf of Mimosa was cut off from the plant, and the subsequent histories of the wounded plant and the detached leaf are curiously different. The cutting of one of its leaves had caused a great shock to the parent plant, and an intense excitation spreads over to the distant organs. All the leaves remained depressed and irresponsive for several hours. From this state of paralysed sensibility, the plant gradually recovers and the leaves begin to show returning sensitiveness. The detached leaf, when placed in a nourishing solution soon recovers, and holds up its head with an attitude indicative of defiance, and the responses it gives are energetic. This lasts for twenty four hours, after which a curious change creeps in the vigour of its responses begins rapidly to wane. The leaf hitherto erect, falls over; death had at last asserted its mastery.

—*Amrita Bazar Patrika*, 10-2-1919.

LIFE AND SPEECHES OF EMINENT INDIANS

The Hon. Pandit Madan Mohan Malaviya. His Life and Speeches. (Second edition, revised and enlarged). 700 pages. Price Rs. 3.

Lokamanya B. G. Tilak. An exhaustive and up to date collection of all the soul stirring speeches of the apostle of Home Rule with a valuable appreciation by Babu Aurobinda Ghose. Second edition, revised and enlarged. Price Rs. 2.

Mahatma Gandhi. His Life, Writings and Speeches with a foreword by Mrs. Sarojini Naidu. (Enlarged and up to date edition). Over 450 pages. Tastefully bound with an index. Price Rs. 2.

Mohomed Ali Jinnah. With a Foreword by the Rajah of Mahmudabad. Over 320 pp. Attractively bound with a portrait and an index. Price Rs. 2.

Babu Surendranath Banerjee. An exhaustive collection of all the speeches of Babu Surendranath Banerjee delivered in England. Price As. 8.

India for Indians. A collection of the speeches delivered by Mr. C. R. Das on Home Rule for India with an Introduction by Babu Motilal Ghose. Second Edition, revised and enlarged. Price As. 12.

Sir Rabindranath Tagore. His Life, Personality, and Genius, by K. S. Ramaswami Sastri, B.A., B.L. with a Foreword by Mr. J. C. Rollo. Price Rs. 3.

J. N. Tata. His Life and Life Work. By Sir D. E. Wacha. 3rd edition. Price Re. 1.

GANESH & CO., PUBLISHERS, MADRAS.

www.ingramcontent.com/pod-product-compliance
Lightning Source LLC
Chambersburg PA
CBHW031624210526
45464CB00004B/1743